AN INTRODUCTION TO GENERALIZED LINEAR MODELS

Annette J. Dobson

University of Newcastle
New South Wales, Australia

CHAPMAN & HALL/CRC

Boca Raton London New York Washington, D.C.

Library of Congress Cataloging-in-Publication Data

Catalog record is available from the Library of Congress.

Direct all inquiries to CRC Press LLC, 2000 N.W. Corporate Blvd., Boca Raton, Florida 33431.

First edition 1990
Reprinted 1991, 1993, 1994, 1996, 1997
First CRC Reprint 1999
Originally published by Chapman & Hall
No claim to original U.S. Government works
International Standard Book Number 0-412-31110-0
Printed in the United States of America 4 5 6 7 8 9 0
Printed on acid-free paper

Contents

Preface

This book is about generalized linear models. These models provide a unified theoretical and conceptual framework for many of the most commonly used statistical methods: simple and multiple regression, t-tests and analysis of variance and covariance, logistic regression, log-linear models for contingency tables and several other analytical methods.

The first edition, *An Introduction to Statistical Modelling*, was written at the same time as the first edition of McCullagh and Nelder's book *Generalized Linear Models* (1983; second edition 1989) and the market position of the two books was uncertain. Also the book appeared before the release of an improved version of GLIM, a statistical program developed to do the calculations required for generalized linear models. GLIM 3.77 with its rewritten manual and the generalized linear modelling programs which are now available in other statistical computing systems make it easier to do numerical exercises which illustrate the theory.

In the light of experience it became apparent that a new edition of the book was needed specifically to provide an introduction to generalized linear models for intermediate or higher level undergraduates and postgraduates. The title should reflect the level and content more accurately. The new edition is intended to fulfil these requirements. A more advanced treatment is given by McCullagh and Nelder (1989).

It is assumed that the reader has a working knowledge of basic statistical concepts and methods (at the level of most introductory statistics courses) and some acquaintance with calculus and matrix algebra.

Emphasis is on the use of statistical models to investigate substantive questions rather than to produce mathematical descriptions of the data. Therefore parameter estimation and hypothesis testing are stressed.

Differences from the first edition are as follows:

1. More detailed explanations have been given in many parts;
2. Several chapters have been extensively rewritten;
3. There are more examples and exercises, particularly numerical ones, and outlines of solutions for selected exercises are given in the back of the book.

I would like to thank everyone who has offered comments, criticisms and suggestions about the first edition. These have been most helpful in preparing the second one. However, the choice of material and the errors and obscurities are my responsibility.

Annette J. Dobson
Department of Statistics
University of Newcastle
New South Wales, Australia

1

Introduction

1.1 BACKGROUND

This book is designed to introduce the reader to the ideas of statistical modelling and, in particular, to the generalized linear model in order to demonstrate the unity among many commonly used statistical techniques. It is assumed that the reader already has some familiarity with statistical principles and methods; in particular, sampling distributions, hypothesis testing, *t*-tests, analysis of variance, simple linear regression and chi-squared tests of independence for two-dimensional contingency tables. In addition, some knowledge of matrix algebra and calculus is assumed.

The reader will find it necessary to have access to statistical computing facilities. In particular it is assumed that the programs **GLIM** (Numerical Algorithms Group, Oxford, UK) and **MINITAB** (Minitab Data Analysis Software, Pennsylvania, USA) are available. Other programs such as **SAS** (SAS Institute, Cary, North Carolina, USA), **SYSTAT** (SYSTAT Inc., Evanston, Illinois, USA), **BMDP** (BMDP Statistical Software, Los Angeles, USA), **SPSS** (SPSS Inc., Chicago, Illinois, USA) or **GENSTAT** (Rothamsted Experimental Station, Herts, UK) may also be useful as they are more comprehensive, at least in some areas, than GLIM or MINITAB.

1.2 SCOPE

The statistical methods considered in this book all involve the analysis of relationships between multiple measurements made on groups of subjects or objects. For example, the measurements might be the heights or weights and the ages of boys and girls, or the yield of plants under various growing conditions. We use the terms **response, outcome** or **dependent variable** for measurements we regard as random variables. These measures are free to vary in response to other variables called **independent, explanatory** or **predictor variables** which are treated as though they are non-random measurements or observations (e.g. those fixed by experimental design).

Measurements are made on one of the following scales.

1. **Nominal** classifications: e.g. red, green, blue; yes, no, do not know, not applicable. In particular, for **binary** or **dichotomous** variables there are only two categories: male, female; dead, alive; smooth leaves, serrated leaves;
2. **Ordinal** classifications in which there is some natural order or ranking between the categories: e.g. young, middle-aged, old; diastolic blood pressures grouped as ≤ 70, 71–90, 91–110, 111–130, ≥ 131 mm Hg;
3. **Continuous** measurements where observations may, at least in theory, fall anywhere on a continuum: e.g. weight, length or time. (This scale includes both **interval** and **ratio scale** measurements – the latter have a well-defined zero.)

Usually nominal and ordinal data are recorded as the numbers of observations in each category. These counts or **frequencies** are called **discrete** variables. For continuous data the individual measurements are recorded. The term **quantitative** is often used for a variable measured on a continuous scale and the term **qualitative** for nominal and sometimes for ordinal measurements. A qualitative, explanatory variable is called a **factor** and its categories are called the **levels** of the factor. A quantitative explanatory variable is called a **covariate**.

Methods of statistical analysis depend on the measurement scales of the response and explanatory variables. In practice ordinal data, because of their intermediate position between nominal and continuous observations, are often analysed by methods designed for one of the other two scales. In this book methods specific to ordinal measurements are rarely considered due to limitations of space rather than as an endorsement of methods which are not strictly appropriate.

Generally we consider only those statistical methods which are relevant when there is just *one response variable* although there will usually be several explanatory variables. For various combinations of response and explanatory variables Table 1.1 shows the main methods of statistical analysis and the chapters in which these are described.

Chapters 2–5 cover the theoretical framework which is common to the subsequent chapters which focus on methods for analysing particular kinds of data.

Chapter 2 develops the idea of statistical modelling via some numerical examples. The modelling process involves three steps:

1. Specifying plausible equations and probability distributions (models) to describe the main features of the response variable;
2. Estimating parameters used in the models;
3. Making inferences; for example, testing hypotheses by considering how adequately the models fit the actual data.

Table 1.1 Major methods of statistical analysis for response and explanatory variables measured on various scales

Explanatory variables	Response variable: Binary	Response variable: Nominal with more than two categories	Response variable: Continuous
Binary	2 × 2 contingency tables, logistic regression (Chapter 8), and log-linear models (Chapter 9)	Contingency tables and log-linear models (Chapter 9)	*t*-tests (Chapter 7)
Nominal with more than two categories	Generalized logistic regression (Chapter 8) and log-linear models (Chapter 9)	Contingency tables and log-linear models (Chapter 9)	Analysis of variance (Chapter 7)
Continuous	Dose–response models including logistic regression (Chapter 8)	[a]	Multiple regression (Chapter 6)
Some continuous and some categorical	Generalized logistic regression models (Chapter 8)	[a]	Analysis of covariance (Chapter 7) and multiple regression (Chapter 6)

[a]Data involving a nominal response variable with more than two categories and continuous explanatory variables are often analysed by redefining the problem so that the roles of the response and explanatory variables are interchanged.

In the numerical examples the modelling approach is compared with more traditional analyses of the same data sets.

The next three chapters concern the theoretical development of each of the three steps just outlined. Chapter 3 is about the exponential family of distributions, which includes the Normal, Poisson and binomial distributions. It also covers generalized linear models (as defined by Nelder and Wedderburn, 1972) of which linear regression and many other models are special cases. In Chapter 4 two methods of estimation, maximum likelihood and least squares, are considered. For some generalized linear models these methods give identical results but for others likelihood methods are often more useful. Chapter 5 concerns test statistics which provide measures of how well a model describes a given set of data. Hypothesis testing is carried out by first specifying alternative models (one corresponding to the null hypothesis and the other to a more general hypothesis). Then test statistics are calculated which measure the 'goodness of fit' of each model. Finally the test statistics are compared. Typically the model corresponding to the null hypothesis is simpler, so if it fits the data approximately as well as the other model we usually prefer it on the grounds of parsimony (i.e. we retain the null hypothesis).

Chapter 6 is about multiple linear regression which is the standard method for relating a continuous response variable to several continuous explanatory (or predictor) variables. Analysis of variance (ANOVA) and analysis of covariance (ANCOVA) are discussed in Chapter 7. ANOVA is used for a continuous response variable and qualitative explanatory variables (factors). ANCOVA is used when at least one of the explanatory variables is qualitative and at least one is continuous. This distinction between multiple linear regression and ANCOVA (or even ANOVA) is somewhat artificial. The methods are so closely related that nowadays it is common to use the same computational tools for all such problems. The terms multiple regression or **general linear model** are used to cover the spectrum of methods for analysing one continuous response variable and multiple explanatory variables.

Chapter 8 is about methods for analysing binary response data. The most common one is logistic regression which is used to model relationships between the response variable and several explanatory variables which may be categorical or continuous. Methods for relating the response to a single continuous variable, the dose, are also considered; these include probit analysis which was originally developed for analysing dose-response data from bioassays.

Chapter 9 concerns contingency tables and is mainly about log-linear models which are used to investigate relationships between several categorical variables. In this chapter the distinction between response and explanatory variables is less crucial and the methods are also suitable for multiple responses.

Other statistical methods which can fit into the same general framework include Poisson regression, certain models for survival time data and the Bradley–Terry model for paired comparisons. Further examples of generalized linear models are discussed in the books by McCullagh and Nelder (1989), Andersen (1980), Aitkin, Anderson, Francis and Hinde (1989) and Healy (1988).

1.3 NOTATION

Generally we follow the convention of denoting random variables by upper-case italic letters and observed values by the corresponding lower-case letters; for example, the observations $y_1, y_2, \ldots, y_N$ are regarded as realizations of the random variables $Y_1, Y_2, \ldots, Y_N$. Greek letters are used to denote parameters and the corresponding lower-case italic letters to denote estimators and estimates; occasionally the symbol $\hat{}$ is used for estimators or estimates. For example, the parameter β is estimated by $\hat{\beta}$ or b. Sometimes these conventions are not strictly adhered to, either to avoid excessive notation in cases when the meaning should be apparent from the context, or when there is a strong tradition of alternative notation (e.g. e or ε for random error terms).

Vectors and matrices, whether random or not, are denoted by bold-face lower-case roman and bold-face italic upper-case letters respectively; for example, $\mathbf{y}$ represents a vector of observations

$$\begin{bmatrix} y_1 \\ \vdots \\ y_N \end{bmatrix}$$

or a vector of random variables

$$\begin{bmatrix} Y_1 \\ \vdots \\ Y_N \end{bmatrix}$$

$\boldsymbol{\beta}$ denotes a vector of parameters and $\boldsymbol{X}$ is a matrix. The superscript T is used for matrix transpose or when a column vector is written as a row, e.g. $\mathbf{y} = [y_1, \ldots, y_N]^{\mathrm{T}}$.

The probability density function of a continuous random variable Y (or the probability distribution if Y is discrete) is denoted by

$$f(y; \boldsymbol{\theta})$$

where $\boldsymbol{\theta}$ represents the parameters of the distribution.

We use dot (.) subscripts for summation and bars ($^{-}$) for means, thus

$$\bar{y} = \frac{1}{N}\sum_{i=1}^{N} y_i = \frac{1}{N}\, y.$$

1.4 DISTRIBUTIONS DERIVED FROM THE NORMAL DISTRIBUTION

The sampling distributions of many of the statistics used in this book depend on the Normal distribution. They do so either directly, because they are based on Normally distributed random variables, or asymptotically, via the central limit theorem for large samples. In this section we give definitions and notation for these distributions and summarize the relationships between them. The exercises at the end of the chapter provide practice in using these results which are employed extensively in subsequent chapters.

1.4.1 Normal distributions

1. If the random variable Y has the **Normal distribution** with mean μ and variance σ^2 we denote this by $Y \sim N(\mu, \sigma^2)$;
2. The Normal distribution with $\mu = 0$ and $\sigma^2 = 1$, that is $Y \sim N(0, 1)$, is called the **standard Normal distribution**;
3. Let $Y_1, \ldots, Y_n$ denote Normally distributed random variables with $Y_i \sim N(\mu_i, \sigma_{ii}^2)$ for $i = 1, \ldots, n$ and let the covariance of Y_i and Y_j be denoted by $\text{cov}(Y_i, Y_j) = \rho_{ij}\sigma_i\sigma_j$. Then the joint distribution of the Y's is the **multivariate Normal distribution** with mean vector $\boldsymbol{\mu} = [\mu_1, \ldots, \mu_n]^{\mathrm{T}}$ and variance–covariance matrix $\boldsymbol{V}$ which has elements $\rho_{ij}\sigma_i\sigma_j$.. We write this as $\mathbf{y} \sim N(\boldsymbol{\mu}, \boldsymbol{V})$ where $\mathbf{y} = [Y_1, \ldots, Y_n]^{\mathrm{T}}$;
4. Suppose the random variables $Y_1, \ldots, Y_n$ are independent and Normally distributed with the distributions $Y_i \sim N(\mu_i, \sigma_i^2)$ for $i = 1, \ldots, n$. Suppose that the random variable W is a linear combination of the Y's

$$W = a_1Y_1 + a_2Y_2 + \ldots + a_nY_n$$

where the a_i's are constants. Then the mean (or expected value) of W is

$$E(W) = a_1\mu_1 + a_2\mu_2 + \ldots + a_n\mu_n$$

and its variance is

$$\text{var}(W) = a_1^2\sigma_1^2 + a_2^2\sigma_2^2 + \ldots + a_n^2\sigma_n^2$$

Furthermore W is Normally distributed, so that

$$W = \sum_{i=1}^{n} a_iY_i \sim N\left(\sum_{i=1}^{n} a_i\mu_i, \sum_{i=1}^{n} a_i^2\sigma_i^2\right)$$

1.4.2 Chi-squared distributions

1. The **central chi-squared distribution** with n **degrees of freedom** is defined as the sum of squares of n independent random variables $Z_1, \ldots, Z_n$ each with the standard Normal distribution. It is denoted by

$$X^2 = \sum_{i=1}^{n} Z_i^2 \sim \chi_n^2$$

In matrix notation this is $X^2 = \mathbf{z}^T\mathbf{z} \sim \chi_n^2$ where $\mathbf{z} = [Z_1, \ldots, Z_n]^T$.

2. If $Y_1, \ldots, Y_n$ are independent Normally distributed random variables with the distributions $Y_i \sim N(\mu_i, \sigma_i^2)$ then

$$X^2 = \sum_{i=1}^{n} \left(\frac{Y_i - \mu_i}{\sigma_i} \right)^2 \sim \chi_n^2 \tag{1.1}$$

because the variables $Z_i = (Y_i - \mu_i)/\sigma_i$ have the standard Normal distribution $N(0, 1)$.

3. More generally, suppose that the Y_i's are not necessarily independent and that the vector $\mathbf{y} = [Y_1, \ldots, Y_n]^T$ has the multivariate Normal distribution $\mathbf{y} \sim N(\boldsymbol{\mu}, V)$, where the variance–covariance matrix V is non-singular and its inverse is V^{-1}. Then

$$X^2 = (\mathbf{y} - \boldsymbol{\mu})^T V^{-1} (\mathbf{y} - \boldsymbol{\mu}) \sim \chi_n^2 \tag{1.2}$$

4. If $\mathbf{y} \sim N(\boldsymbol{\mu}, V)$ then the distribution of the random variable $\mathbf{y}^T V^{-1} \mathbf{y}$ is called the **non-central chi-squared distribution** with n degrees of freedom and non-centrality parameters $\lambda = \boldsymbol{\mu}^T V^{-1} \boldsymbol{\mu}/2$. We denote this by

$$\mathbf{y}^T V^{-1} \mathbf{y} \sim \chi^2(n, \lambda)$$

5. If $X_1^2, \ldots, X_m^2$ are independent random variables with the chi-squared distributions $X_i^2 \sim \chi^2(n_i, \lambda_i)$, which may or may not be central, then their sum also has a chi-squared distribution with Σn_i degrees of freedom and non-centrality parameter $\Sigma \lambda_i$, i.e.

$$\sum_{i=1}^{m} X_i^2 \sim \chi^2\left(\sum_{i=1}^{m} n_i, \sum_{i=1}^{m} \lambda_i \right)$$

This is called the **reproductive property** of the chi-squared distribution.

6. A related result, which is used frequently in this book, is that if X_1^2 and X_2^2 have chi-squared distributions with n_1 and n_2 degrees of freedom respectively, where $n_1 > n_2$, and they are independent then their difference also has a chi-squared distribution

$$X_1^2 - X_2^2 \sim \chi_{n_1 - n_2}^2$$

(A similar result holds for non-central chi-squared distributions.)

7. Let $\mathbf{y} \sim N(\boldsymbol{\mu}, V)$ where $\mathbf{y}$ has n elements and V is singular with rank $k < n$ so that the inverse of V is not uniquely defined but let V^- denote a generalized inverse of V. Then the random variable $\mathbf{y}^T V^- \mathbf{y}$ has the chi-squared distribution with k degrees of freedom and non-centrality parameter $\lambda = \boldsymbol{\mu}^T V^- \boldsymbol{\mu}/2$.

1.4.3 *t*-distribution

The ***t*-distribution** with n degrees of freedom is defined as the ratio of two independent random variables, one of which has the standard Normal distribution and the other is the square root of a central chi-squared random variable divided by its degrees of freedom; that is

$$T = \frac{Z}{(X^2/n)^{1/2}} \tag{1.3}$$

where $Z \sim N(0, 1)$, $X^2 \sim \chi^2_n$ and Z and X^2 are independent. This is denoted by $T \sim t_n$.

1.4.4 *F*-distributions

1. The **central *F*-distribution** with n and m degrees of freedom is defined as the ratio of two independent central chi-squared random variables each divided by its degrees of freedom,

$$F = \frac{X_1^2/n}{X_2^2/m} \tag{1.4}$$

where $X_1^2 \sim \chi^2_n$, $X_2^2 \sim \chi^2_m$ and X_1^2 and X_2^2 are independent. This is denoted by $F \sim F_{n,m}$.

2. The relationship between the t-distribution and the F-distribution can be derived by squaring the terms in equation (1.3) and using definition (1.4) to obtain

$$T^2 = \frac{Z^2/1}{X^2/n} \sim F_{1,n} \tag{1.5}$$

That is, the square of a random variable with the t-distribution t_n has the F-distribution $F_{1,n}$.

3. The **non-central *F*-distribution** is defined as the ratio of two independent random variables, each divided by its degrees of freedom, where the numerator has a non-central chi-squared distribution and the denominator has a central chi-squared distribution, i.e.

$$F = \frac{X_1^2/n}{X_2^2/m}$$

where $X_1^2 \sim \chi^2(n, \lambda)$, $X_2^2 \sim \chi^2_m$ and X_1^2 and X_2^2 are independent.

1.5 EXERCISES

1.1 Let Y_1 and Y_2 be independent random variables with $Y_1 \sim N(1, 3)$ and $Y_2 \sim N(2, 5)$. If $W_1 = Y_1 + 2Y_2$ and $W_2 = 4Y_1 - Y_2$ what is the joint distribution of W_1 and W_2?

1.2 Let Y_1 and Y_2 be independent random variables with $Y_1 \sim N(0, 1)$ and $Y_2 \sim N(3, 4)$.

(a) What is the distribution of Y_1^2?
(b) If

$$\mathbf{y} = \begin{bmatrix} Y_1 \\ (Y_2 - 3)/2 \end{bmatrix}$$

what is the distribution of $\mathbf{y}^T\mathbf{y}$?
(c) If

$$\mathbf{y} = \begin{bmatrix} Y_1 \\ Y_2 \end{bmatrix}$$

and its distribution is denoted by $\mathbf{y} \sim N(\boldsymbol{\mu}, \boldsymbol{V})$ what is the distribution of $\mathbf{y}^T\boldsymbol{V}^{-1}\mathbf{y}$?

1.3 If $Y_1, \ldots, Y_n$ are a random sample of observations from the distribution $N(\mu, \sigma^2)$, then it is well known that

$$\bar{Y} = \frac{1}{n}\sum_{i=1}^{n} Y_i \quad \text{and} \quad S^2 = \frac{1}{n-1}\sum_{i=1}^{n}(Y_i - \bar{Y})^2$$

are independent; proofs can be found in many elemenatry textbooks.

(a) What is the distribution of $\bar{Y}$?
(b) Show that

$$S^2 = \frac{1}{n-1}\left[\sum_{i=1}^{n}(Y_i - \mu)^2 - n(\bar{Y} - \mu)^2\right]$$

(c) What is the distribution of $(n-1)S^2/\sigma^2$?
(d) What is the distribution of

$$\frac{\bar{Y} - \mu}{S/\sqrt{n}}?$$

2

Model fitting

2.1 INTRODUCTION

The transmission and reception of information involves a message, or **signal**, which is distorted by **noise**. It is sometimes useful to think of scientific data as measurements composed of signal and noise and to construct mathematical models incorporating both of these components. Often the signal is regarded as **deterministic** (i.e. non-random) and the noise as random. Therefore, a mathematical model of the data combining both signal and noise is probabilistic and it is called a statistical model.

Another way of thinking of a statistical model is to consider the signal component as a mathematical description of the main features of the data and the noise component as all those characteristics not 'explained' by the model (i.e. by its signal component).

Our goal is to extract from the data as much information as possible about the signal. The first step is to postulate a model, in the form of an equation involving the signal and noise and a probability distribution describing the form of random variation. Typically the mathematical description of the signal involves several unknown constants, termed **parameters**. The next step is to estimate values for the parameters from the data.

Once the signal component has been quantified we can partition the total variability observed in the data into a portion attributable to the signal and the remainder attributable to the noise. A criterion for a good model is one which 'explains' a large proportion of this variability, i.e. one in which the part attributable to signal is large relative to the part attributable to noise. In practice, this has to be balanced against other criteria such as simplicity. The *Oxford English Dictionary* describes the law of parsimony (otherwise known as Occam's Razor) as the principle that no more causes should be assumed than will account for the effect. According to this principle a simpler model which describes the data adequately (i.e. a parsimonious model) may be preferable to a more complicated one which leaves little of the variability 'unexplained'.

Often we wish to test hypotheses about the parameters. This can be performed in the context of model fitting by defining a series of models corresponding to different hypotheses. Then the question about whether the data support a particular hypothesis can be formulated in terms of the adequacy of fit of the corresponding model (i.e. the amount of variability it explains) relative to other models.

These ideas are now illustrated by two detailed examples.

2.2 PLANT GROWTH EXAMPLE

Suppose that genetically similar seeds are randomly assigned to be raised either in a nutritionally enriched environment (treatment) or under standard conditions (control) using a **completely randomized experimental design**. After a predetermined period all plants are harvested, dried and weighed. The results, expressed as dried weight in grams, for samples of ten plants from each environment are given in Table 2.1. Figure 2.1 shows a dot plot of the distributions of these weights.

Table 2.1 Plant weights from two different growing conditions

Control (1)	4.17	5.58	5.18	6.11	4.50	4.61	5.17	4.53	5.33	5.14
Treatment (2)	4.81	4.17	4.41	3.59	5.87	3.83	6.03	4.89	4.32	4.69

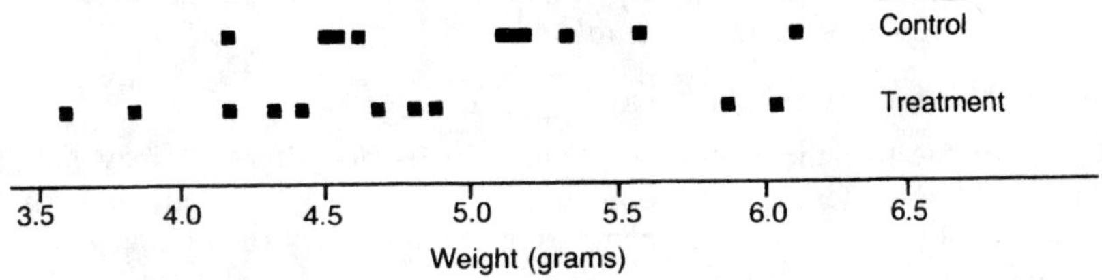

Figure 2.1 Plant growth data from Table 2.1.

The first step is to formulate models to describe these data, for example

Model 1: $$Y_{jk} = \mu_j + e_{jk} \tag{2.1}$$

where Y_{jk} is the weight of the kth plant ($k = 1, \ldots, K$ with $K = 10$ in this case) from the jth sample (with $j = 1$ for control and $j = 2$ for treatment);

μ_j is a parameter, the signal component of weight, determined by the growth environment. It represents a common characteristic of all plants grown under the conditions experienced by sample j;

e_{jk} is the noise component. It is a random variable (although by convention it is usually written using lower case). It is sometimes called the **random error term**. It represents that element of weight unique to the kth observation from sample j.

From the design of the experiment we assume that the e_{jk}'s are independent. We also assume that they are identically distributed with the Normal distribution with mean zero and variance σ^2, i.e. $e_{jk} \sim N(0, \sigma^2)$. Therefore the Y_{jk}'s are also independent and $Y_{jk} \sim N(\mu_j, \sigma^2)$ for all j and k.

We would like to know if the enriched environment made a difference to the weight of the plants so we need to estimate the difference between μ_1 and μ_2 and test whether it differs significantly from some pre-specified value (such as zero).

An alternative specification of the model which is more suitable for comparative use is

Model 2: $$Y_{jk} = \mu + \alpha_j + e_{jk} \qquad (2.2)$$

where Y_{jk} and e_{jk} are defined as before;

μ is a parameter representing that aspect of growth common to both environments;

α_1 and α_2 are parameters representing the differential effects due to the control treatment conditions; formally $\alpha_j = \mu_j - \mu$.

If the nutritionally enriched conditions do not enhance (or inhibit) plant growth, then the terms α_j will be negligible and so the Model 2 (equation 2.2) will be equivalent to

Model 0: $$Y_{jk} = \mu + e_{jk} \qquad (2.3)$$

Therefore, testing the hypothesis that there is no difference in weight due to the different environments (i.e. $\mu_1 = \mu_2$ or equivalently $\alpha_1 = \alpha_2 = 0$) is equivalent to comparing the adequacy of Models 1 and 0 (equations 2.1 and 2.3) for describing the data.

The next step is to estimate the model parameters. We will do this using the **likelihood function**. This is the same as the joint probability density function of the response variables Y_{jk}, but whereas the joint probability density function is regarded as a function of the random variables Y_{jk} (conditional on the parameters), the likelihood function is viewed primarily as a function of the parameters, conditional on the observations y_{jk}. **Maximum likelihood estimators** are the values of the parameters which correspond to the maximum value of the likelihood function, or equivalently, to the maximum of the logarithm of the likelihood function which is called the **log-likelihood function**.

We begin by estimating parameters μ_1 and μ_2 in Model 1 (equation 2.1), treating σ^2 as a known constant (in this context σ^2 is often referred to as a **nuisance parameter**). Since the Y_{jk}'s are independent, the likelihood function is the product of their probability density functions

$$\prod_{j=1}^{2}\prod_{k=1}^{K}\frac{1}{(2\pi\sigma^2)^{1/2}}\exp\left\{-\frac{1}{2\sigma^2}(y_{jk}-\mu_j)^2\right\}$$

and so the log-likelihood function is

$$l_1 = -K\log(2\pi\sigma^2) - \frac{1}{2\sigma^2}\sum_{j=1}^{2}\sum_{k=1}^{K}(y_{jk}-\mu_j)^2$$

The maximum likelihood estimators of μ_1 and μ_2 are obtained by solving the simultaneous equations

$$\frac{\partial l_1}{\partial \mu_j} = \frac{1}{\sigma^2}\sum_{k=1}^{K}(y_{jk}-\mu_j) = 0, \qquad j = 1, 2$$

So the estimators are given by

$$\frac{1}{\sigma^2}\sum_{k=1}^{K} y_{jk} = \frac{1}{\sigma^2}K\hat{\mu}_j$$

Hence

$$\hat{\mu}_j = \frac{1}{K}\sum_{k=1}^{K} y_{jk} = \bar{y}_j \qquad \text{for } j = 1, 2$$

By considering the second derivatives it can be verified that $\hat{\mu}_1$ and $\hat{\mu}_2$ do, in fact, correspond to the maximum of l_1.

Thus the maximum value of l_1, denoted by $\hat{l}_1$, is given by

$$\hat{l}_1 = -K\log(2\pi\sigma^2) - \frac{1}{2\sigma^2}\hat{S}_1$$

where

$$\hat{S}_1 = \sum_{j=1}^{2}\sum_{k=1}^{K}(y_{jk}-\bar{y}_j)^2$$

Now we consider Model 0 (equation 2.3) and again find the maximum likelihood estimators and the maximum value of the log-likelihood function. The likelihood function is

$$\prod_{j=1}^{2}\prod_{k=1}^{K}\frac{1}{(2\pi\sigma^2)^{1/2}}\exp\left\{-\frac{1}{2\sigma^2}(y_{jk}-\mu)^2\right\}$$

because the Y_{jk}'s are independent and all have the same distribution $N(\mu, \sigma^2)$. Therefore the log-likelihood function is

$$l_0 = -K\log(2\pi\sigma^2) - \frac{1}{2\sigma^2}\sum_{j=1}^{2}\sum_{k=1}^{K}(y_{jk}-\mu)^2$$

The maximum likelihood estimator $\hat{\mu}$ is the solution of the equation

$\partial l_0/\partial\mu = 0$, that is,

$$\hat{\mu} = \frac{1}{2K}\sum_{j=1}^{2}\sum_{k=1}^{K} y_{jk} = \bar{y}$$

Therefore the maximum value attained by the log-likelihood function is

$$\hat{l}_0 = -K\log(2\pi\sigma^2) - \frac{1}{2\sigma^2}\hat{S}_0$$

where

$$\hat{S}_0 = \sum_{j=1}^{2}\sum_{k=1}^{K}(y_{jk} - \bar{y})^2$$

For the plant data the values of the maximum likelihood estimates and the statistics $\hat{S}_1$ and $\hat{S}_0$ are shown in Table 2.2.

Table 2.2 Analysis of plant growth data in Table 2.1

Model 1	$\hat{\mu}_1 = 5.032$, $\hat{\mu}_2 = 4.661$ and $\hat{S}_1 = 8.729$
Model 0	$\hat{\mu} = 4.8465$ and $\hat{S}_0 = 9.417$

The third step in the model-fitting procedure involves testing hypotheses. If the null hypothesis

$$H_0 : \mu_1 = \mu_2$$

is correct then Model 1 and Model 0 are the same so the maximum values $\hat{l}_1$ and $\hat{l}_0$ of the log-likelihood functions should be nearly equal, or equivalently, $\hat{S}_1$ and $\hat{S}_0$ should be nearly equal. If the data support this hypothesis, we would feel justified in using the simpler Model 0 to describe the data. On the other hand, if the more general hypothesis

$$H_1 : \mu_1 \text{ and } \mu_2 \text{ are not necessarily equal}$$

is true then $\hat{S}_0$ should be larger than $\hat{S}_1$ (corresponding to $\hat{l}_0$ smaller than $\hat{l}_1$) and Model 1 would be preferable.

To assess the relative magnitude of $\hat{S}_1$ and $\hat{S}_0$ we need to consider the sampling distributions of the corresponding random variables

$$S_1 = \sum_{j=1}^{2}\sum_{k=1}^{K}(Y_{jk} - \bar{Y}_j)^2 \quad \text{and} \quad S_0 = \sum_{j=1}^{2}\sum_{k=1}^{K}(Y_{jk} - \bar{Y})^2$$

It can be shown (as in Exercise 1.3(b)) that

$$\frac{1}{\sigma^2}S_1 = \frac{1}{\sigma^2}\sum_{j=1}^{2}\sum_{k=1}^{K}(Y_{jk} - \bar{Y}_j)^2$$

$$= \frac{1}{\sigma^2} \sum_{j=1}^{2} \sum_{k=1}^{K} (Y_{jk} - \mu_j)^2 - \frac{K}{\sigma^2} \sum_{j=1}^{2} (\bar{Y}_j - \mu_j)^2$$

For the more general Model 1 if the Y_{jk}'s are independent with the distributions $N(\mu_j, \sigma^2)$ then the group means $\bar{Y}_j$ will also be independent with the distributions $\bar{Y}_j \sim N(\mu_j, \sigma^2/K)$. Therefore (S_1/σ^2) is the difference between the sum of the squares of $2K$ independent random variables $(Y_{jk} - \mu_j)/\sigma$ which each have the distribution $N(0, 1)$ and the sum of squares of two independent random variables $(\bar{Y}_j - \mu_j)/(\sigma^2/K)^{1/2}$ which also have the $N(0, 1)$ distribution. Hence, from the properties of the chi-squared distribution (Section 1.4.2)

$$\frac{1}{\sigma^2} S_1 \sim \chi^2_{2K-2}$$

For the simpler Model 0 a similar argument applies. Let $\bar{\mu} = (\mu_1 + \mu_2)/2$

$$\frac{1}{\sigma^2} S_0 = \frac{1}{\sigma^2} \sum_{j=1}^{2} \sum_{k=1}^{K} (Y_{jk} - \bar{Y})^2$$

$$= \frac{1}{\sigma^2} \sum_{j=1}^{2} \sum_{k=1}^{K} (Y_{jk} - \bar{\mu})^2 - \frac{2K}{\sigma^2} (\bar{Y} - \bar{\mu})^2$$

Since the Y_{jk}'s are assumed to be independent and to have the distributions $N(\mu_j, \sigma^2)$ their mean $\bar{Y}$ has the distribution $N(\bar{\mu}, \sigma^2/2K)$. So the second term in the expression for (S_0/σ^2) is the square of one random variable with the $N(0, 1)$ distribution. Also if $\mu_1 = \mu_2 = \bar{\mu}$ (corresponding to H_0) then the first term of (S_0/σ^2) is the sum of the squares of $2K$ independent random variables $(Y_{jk} - \bar{\mu})/\sigma$ each with the distribution $N(0, 1)$. Therefore from the properties of the chi-squared distribution

$$\frac{1}{\sigma^2} S_0 \sim \chi^2_{2K-1}$$

However, if μ_1 and μ_2 are not necessarily equal (corresponding to H_1) then $(Y_{jk} - \bar{\mu})/\sigma$ has the distribution $N(\mu_j - \bar{\mu}, 1)$ so that (S_0/σ^2) has a non-central chi-squared distribution with $2K - 1$ degrees of freedom.

The statistic $S_0 - S_1$ represents the difference in fit between the two models. If $H_0 : \mu_1 = \mu_2$ is correct then

$$\frac{1}{\sigma^2} (S_0 - S_1) \sim \chi^2_1$$

otherwise it has a non-central chi-squared distribution. However, since σ^2 is unknown we cannot compare $S_0 - S_1$ directly with the χ^2_1 distribution. Instead we eliminate σ^2 by using the ratio of $(S_0 - S_1)/\sigma^2$ and the central chi-squared random variable (S_1/σ^2), each divided by its degrees of freedom, i.e.

$$F = \frac{(S_0 - S_1)/\sigma^2}{1} \Bigg/ \frac{S_1/\sigma^2}{2K - 2} = \frac{S_0 - S_1}{S_1/(2k - 2)}$$

If H_0 is correct, by definition (1.4), F has the central F-distribution with 1 and $(2K - 2)$ degrees of freedom; otherwise F has a non-central F-distribution and so it is likely to be larger than predicted by the central $F_{1,2K-2}$ distribution.

For the plant weight data

$$f = \frac{9.417 - 8.729}{8.729/18} = 1.42$$

which is not statistically significant when compared with the $F_{1,18}$ distribution. Thus the data do not provide evidence against H_0. So we conclude that there is probably no difference in weight due to the different environmental conditions and we can use the simpler Model 0 (equation 2.3) to describe the data.

The more conventional approach to testing H_0 against H_1 is to use a t-test, i.e. to calculate

$$T = \frac{\bar{Y}_1 - \bar{Y}_2}{s(1/K + 1/K)^{1/2}}$$

where s^2, the pooled variance, is defined as

$$s^2 = \frac{\Sigma_{k=1}^{K}(Y_{1k} - \bar{Y}_1)^2 + \Sigma_{k=1}^{K}(Y_{2k} - \bar{Y}_2)^2}{2K - 2}$$

If H_0 is correct the statistic T has the distribution t_{2K-2}. The relationship between the test statistics T and F is obtained as follows:

$$T^2 = \frac{(\bar{Y}_1 - \bar{Y}_2)^2}{2s^2/K}$$

but $s^2 = S_1/(2K - 2)$ and

$$S_0 - S_1 = \sum_{j=1}^{2} \sum_{k=1}^{K} [(Y_{jk} - \bar{Y})^2 - (Y_{jk} - \bar{Y}_j)^2]$$

which can be simplified to

$$S_0 - S_1 = K(\bar{Y}_1 - \bar{Y}_2)^2/2$$

so that

$$T^2 = \frac{S_0 - S_1}{S_1/(2K - 2)} = F$$

corresponding to the distributional relationship that if $T \sim t_n$ then $T^2 \sim F_{1,n}$ (see result 1.5).

The advantages of using an F-test instead of a t-test are:

1. It can be generalized to test the equality of more than two means;
2. It is more closely related to the general methods considered in this book which involve comparing statistics that measure the 'goodness of fit' of competing models.

2.3 BIRTHWEIGHT EXAMPLE

The data in Table 2.3 are the birthweights (g) and estimated gestational ages (weeks) of twelve male and female babies born in a certain hospital. The mean ages are almost the same for both sexes but the mean birthweight for males is higher than for females. The data are shown in the scatter plot in Fig. 2.2. They suggest a linear trend of birthweight increasing with gestational age. The question of interest is whether the rate of increase is the same for males and females.

Table 2.3 Birthweight and gestational age for male and female babies

	Male		Female	
	Age (weeks)	Birthweight (g)	Age (weeks)	Birthweight (g)
	40	2968	40	3317
	38	2795	36	2729
	40	3163	40	2935
	35	2925	38	2754
	36	2625	42	3210
	37	2847	39	2817
	41	3292	40	3126
	40	3473	37	2539
	37	2628	36	2412
	38	3176	38	2991
	40	3421	39	2875
	38	2975	40	3231
Means	38.33	3024.00	38.75	2911.33

A fairly general statistical model for these data is

Model 1: $$Y_{jk} = \alpha_j + \beta_j x_{jk} + e_{jk} \tag{2.4}$$

where the response Y_{jk} is the birthweight for the kth baby of sex j where $j = 1$ for males, $j = 2$ for females and $k = 1, \ldots, K = 12$;

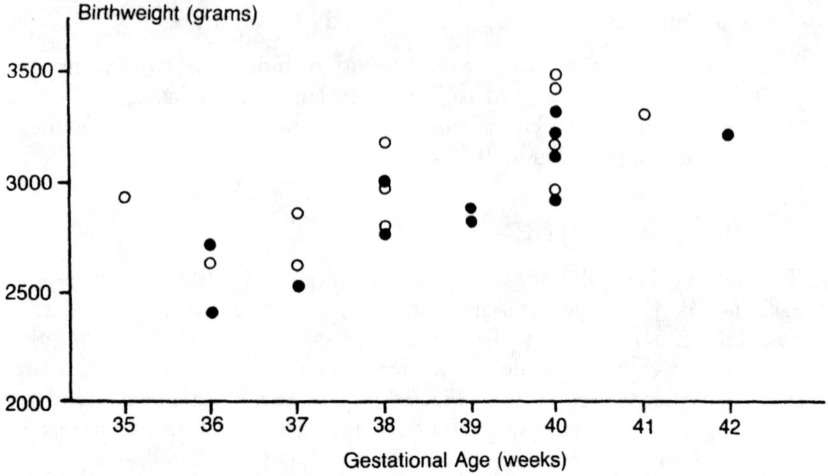

Figure 2.2 Birthweight and gestational age for male and female babies: (○) for males; (●) for females.

the parameters α_1 and α_2 represent the intercepts of the lines for the two sexes;
the parameters β_1 and β_2 represent the slopes or rates of increase of birthweight with age for the two sexes;
the independent variable x_{jk} is the age of the (j, k)th baby (it is not a random variable);
the random error term is e_{jk}; we assume that the e_{jk}'s are independent and that they all have the same distribution $e_{jk} \sim N(0, \sigma^2)$.

If the rate of increase is the same for males and females then a simpler model,

Model 0: $$Y_{jk} = \alpha_j' + \beta x_{jk} + e_{jk} \tag{2.5}$$

is appropriate, where the single parameter β in Model 0 corresponds to the two parameters β_1 and β_2 in Model 1. Thus we can test the null hypothesis

$$H_0: \beta_1 = \beta_2 \ (= \beta)$$

against the more general hypothesis

$$H_1: \beta_1 \text{ and } \beta_2 \text{ not necessarily equal,}$$

by comparing how well Models 1 and 0 fit the data.

The next step in the modelling process is to estimate the parameters. For this example we will use the **method of least squares** instead of the method of maximum likelihood. It consists of minimizing the sum of squares of the differences between the responses and their expected values. For Model 1 (equation 2.4) the expected value of the response Y_{jk} is $E(Y_{jk}) = \alpha_j + \beta_j x_{jk}$ because we assumed that the expected value of the error term e_{jk} was zero.

Therefore the sum of squares to be minimized is

$$S = \sum_j \sum_k (Y_{jk} - \alpha_j - \beta_j x_{jk})^2$$

Geometrically, S is the sum of squares of the vertical distances from the points (x_{jk}, y_{jk}) to the line $y = \alpha_j + \beta_j x$ (see Fig. 2.3). Algebraically it is the sum of squares of the error terms,

$$S = \sum_j \sum_k e_{jk}^2$$

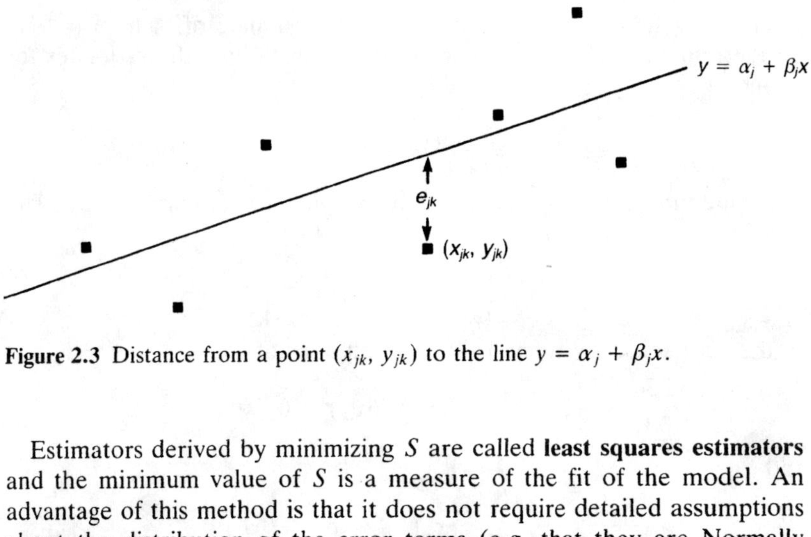

Figure 2.3 Distance from a point (x_{jk}, y_{jk}) to the line $y = \alpha_j + \beta_j x$.

Estimators derived by minimizing S are called **least squares estimators** and the minimum value of S is a measure of the fit of the model. An advantage of this method is that it does not require detailed assumptions about the distribution of the error terms (e.g. that they are Normally distributed). However, such assumptions are required later in order to compare minimum values of S obtained from different models.

First,

Model 1 $$S_1 = \sum_{j=1}^{2} \sum_{k=1}^{K} (Y_{jk} - \alpha_j - \beta_j x_{jk})^2$$

so the least squares estimators for the parameters are the solutions of

$$\frac{\partial S_1}{\partial \alpha_j} = -2 \sum_{k=1}^{K} (Y_{jk} - \alpha_j - \beta_j x_{jk}) = 0$$

$$\frac{\partial S_1}{\partial \beta_j} = -2 \sum_{k=1}^{K} x_{jk}(Y_{jk} - \alpha_j - \beta_j x_{jk}) = 0, \qquad \text{for } j = 1, 2$$

These equations can be simplified to the form

$$\left.\begin{aligned} \sum_{k=1}^{K} Y_{jk} - K\alpha_j - \beta_j \sum_{k=1}^{K} x_{jk} &= 0 \\ \sum_{k=1}^{K} x_{jk}Y_{jk} - \alpha_j \sum_{k=1}^{K} x_{jk} - \beta_j \sum_{k=1}^{K} x_{jk}^2 &= 0 \end{aligned}\right\} \quad j = 1, 2$$

In this form they are called the **Normal equations**. The solutions are

$$b_j = \frac{K\Sigma_k x_{jk}y_{jk} - (\Sigma_k x_{jk})(\Sigma_k y_{jk})}{K\Sigma_k x_{jk}^2 - (\Sigma_k x_{jk})^2}$$

$$a_j = \bar{y}_j - b_j\bar{x}_j$$

where a_j is the estimate of α_j and b_j is the estimate of β_j for $j = 1, 2$. So the minimum value for S_1 obtained by substituting the estimates for a_j and b_j is

$$\hat{S}_1 = \sum_{j=1}^{2} \sum_{k=1}^{K} (y_{jk} - a_j - b_j x_{jk})^2$$

Second, the procedure is repeated for Model 0 (equation 2.5). The expression to be minimized is

$$S_0 = \sum_{j=1}^{2} \sum_{k=1}^{K} (Y_{jk} - \alpha_j - \beta x_{jk})^2$$

so the least squares estimators are the solutions of

$$\frac{\partial S_0}{\partial \alpha_j} = -2 \sum_{k} (Y_{jk} - \alpha_j - \beta x_{jk}) = 0, \qquad j = 1, 2$$

and

$$\frac{\partial S_0}{\partial \beta} = -2 \sum_{j} \sum_{k} x_{jk}(Y_{jk} - \alpha_j - \beta x_{jk}) = 0$$

These are given by the following equations:

$$b = \frac{K\Sigma_j \Sigma_k x_{jk}y_{jk} - \Sigma_j (\Sigma_k x_{jk} \Sigma_k y_{jk})}{K\Sigma_j \Sigma_k x_{jk}^2 - \Sigma_j (\Sigma_k x_{jk})^2}$$

and

$$a_j = \bar{y}_j - b\bar{x}_j$$

For the birthweight example, the data are summarized in Table 2.4. The least squares estimates for both models are given in Table 2.5.

Table 2.4 Summary of birthweight data in Table 2.3 (summation is over $k = 1, \ldots, K$ where $K = 12$)

	Male, $j = 1$	Female, $j = 2$
$\sum x$	460	465
$\sum y$	36 288	34 936
$\sum x^2$	17 672	18 055
$\sum y^2$	110 623 496	102 575 468
$\sum xy$	1 395 370	1 358 497

Table 2.5 Analysis of birthweight data in Table 2.3

Model 1	$b_1 = 111.983$	$a_1 = -1268.672$	
	$b_2 = 130.400$	$a_2 = -2141.667$	$\hat{S}_1 = 652\,424.5$
Model 0	$b = 120.894$	$a_1 = -1610.283$	
		$a_2 = -1773.322$	$\hat{S}_0 = 658\,770.8$

To test the hypothesis $H_0 : \beta_1 = \beta_2$, that is, to compare Models 1 and 0 (equations 2.4 and 2.5 respectively), we need to know the sampling distribution of the minima of the sums of squares. By analogous arguments to those used in the previous example, it can be shown that $(S_1/\sigma^2) \sim \chi^2_{2K-4}$ and if H_0 is correct then $(S_0/\sigma^2) \sim \chi^2_{2K-3}$. In each case the number of degrees of freedom is the number of observations minus the number of parameters estimated. The improvement in fit for Model 1 compared with Model 0 is

$$\frac{1}{\sigma^2}(S_0 - S_1)$$

This can be compared with the fit of the more complicated model 1, that is with (S_1/σ^2), using the test statistic

$$F = \frac{(S_0 - S_1)/1}{S_1/(2K - 4)}$$

If the hypothesis H_0 is correct, $F \sim F_{1,2K-4}$. For these data the value of F is f = 0.2 which is certainly not statistically significant, so the data do not provide evidence against the hypothesis $\beta_1 = \beta_2$ and we have reason for preferring the simpler Model 0.

2.4 NOTATION FOR LINEAR MODELS

The models considered in the above examples can be written in matrix notation in the form

$$\mathbf{y} = X\boldsymbol{\beta} + \mathbf{e} \tag{2.6}$$

where $\mathbf{y}$ is a vector of responses;
$\boldsymbol{\beta}$ is a vector of parameters;
X is a matrix whose elements are zeros or ones or values of 'independent' variables; and
$\mathbf{e}$ is a vector of random error terms.

For quantitative explanatory variables (e.g. age in the birthweight example) the model contains terms of the form βx where the parameter β represents the rate of change in the response corresponding to changes in the independent variable x.

For qualitative explanatory variables there is a parameter to represent each level of a factor (e.g. the effects due to environmental conditions in the plant growth example). The corresponding elements of X are chosen to exclude or include the appropriate parameters for each observation; they are called **dummy variables**. (If only zeros and ones are used for X the term **indicator variable** is used.) X is often called the **design matrix**.

2.5 EXAMPLES

Example 2.1
For the plant growth example Model 1 (equation 2.1) was

$$Y_{jk} = \mu_j + e_{jk} \qquad j = 1, 2 \quad \text{and} \quad k = 1, \ldots, K$$

The corresponding elements of the equation $\mathbf{y} = X\boldsymbol{\beta} + \mathbf{e}$ are

$$\mathbf{y} = \begin{bmatrix} Y_{11} \\ Y_{12} \\ \vdots \\ Y_{1K} \\ Y_{21} \\ \vdots \\ Y_{2K} \end{bmatrix} \quad \boldsymbol{\beta} = \begin{bmatrix} \mu_1 \\ \mu_2 \end{bmatrix} \quad X = \begin{bmatrix} 1 & 0 \\ 1 & 0 \\ \vdots & \vdots \\ 1 & 0 \\ 0 & 1 \\ \vdots & \vdots \\ 0 & 1 \end{bmatrix} \quad \text{and} \quad \mathbf{e} = \begin{bmatrix} e_{11} \\ e_{12} \\ \vdots \\ e_{1K} \\ e_{21} \\ \vdots \\ e_{2K} \end{bmatrix}$$

Example 2.2
For plant growth, the simpler Model 0 (equation 2.3) was

$$Y_{jk} = \mu + e_{jk} \qquad j = 1, 2 \quad \text{and} \quad k = 1, \ldots, K$$

so

$$\mathbf{y} = \begin{bmatrix} Y_{11} \\ Y_{12} \\ \vdots \\ Y_{1K} \\ Y_{21} \\ \vdots \\ Y_{2K} \end{bmatrix} \quad \boldsymbol{\beta} = [\mu] \quad \boldsymbol{X} = \begin{bmatrix} 1 \\ 1 \\ \vdots \\ \vdots \\ 1 \end{bmatrix} \quad \text{and} \quad \mathbf{e} = \begin{bmatrix} e_{11} \\ e_{12} \\ \vdots \\ e_{1K} \\ e_{21} \\ \vdots \\ e_{2K} \end{bmatrix}$$

Example 2.3
For the model

$$Y_{jk} = \alpha_j + \beta_j x_{jk} + e_{jk} \qquad j = 1, 2 \quad \text{and} \quad k = 1, \ldots, K$$

for birthweight (equation 2.4) the corresponding matrix and vector terms are

$$\mathbf{y} = \begin{bmatrix} Y_{11} \\ Y_{12} \\ \vdots \\ Y_{1K} \\ Y_{21} \\ \vdots \\ Y_{2K} \end{bmatrix} \quad \boldsymbol{\beta} = \begin{bmatrix} \alpha_1 \\ \alpha_2 \\ \beta_1 \\ \beta_2 \end{bmatrix} \quad \boldsymbol{X} = \begin{bmatrix} 1 & 0 & x_{11} & 0 \\ 1 & 0 & x_{12} & 0 \\ \vdots & \vdots & \vdots & \vdots \\ 1 & 0 & x_{1K} & 0 \\ 0 & 1 & 0 & x_{21} \\ \vdots & \vdots & \vdots & \vdots \\ 0 & 1 & 0 & x_{2K} \end{bmatrix}$$

and

$$\mathbf{e} = \begin{bmatrix} e_{11} \\ e_{12} \\ \vdots \\ e_{1K} \\ e_{21} \\ \vdots \\ e_{2K} \end{bmatrix}$$

Models of the form $\mathbf{y} = \boldsymbol{X\beta} + \mathbf{e}$ are called **linear models** because the signal part of the model, $\boldsymbol{X\beta}$, is a linear combination of the parameters and the noise part, $\mathbf{e}$, is also additive. If there are p parameters in the model and N observations, then $\mathbf{y}$ and $\mathbf{e}$ are $N \times 1$ random vectors, $\boldsymbol{\beta}$ is a $p \times 1$ vector of parameters (usually to be estimated) and $\boldsymbol{X}$ is an $N \times p$ matrix of known constants.

2.6 EXERCISES

2.1 Use some statistical computing program (e.g. MINITAB) to test the hypothesis that the means under the two growing conditions in the plant weight example (Table 2.1) are equal. (Use a two-sample t-test and, possibly also, a one-way analysis of variance with two groups.) Compare the output with the analysis in the text.

2.2 For the plant growth example find the least squares estimators for the parameters in Model 0 (equation 2.3). Check that they are the same as the maximum likelihood estimators.

2.3 For the data on birthweight (Table 2.3) use some statisical computing program to fit two regression lines, one for each sex, for birthweight against age. Compare the output with the results given in Table 2.5 for Model 1 (equation 2.4). Many standard programs do not allow you to fit the two regression lines with the same slope but different intercepts, as in Model 0 (equation 2.5). Instead fit a single regression line to all the data (for both sexes) and compare the output with the results in Table 2.5.

2.4 The weights (kg) of ten people before and after going on a high carbohydrate diet for three months are shown in Table 2.6. You want to know if, overall, there was any significant change in weight.

Table 2.6 Weights (kg) of ten people before and after a diet

Before	64	71	64	69	76	53	52	72	79	68
After	61	72	63	67	72	49	54	72	74	66

(a) Let Y_{jk} denote the weight of the kth person at time j where $j = 1$ before the diet, $j = 2$ afterwards and $k = 1, \ldots, 10$. Let

$$Y_{jk} = \mu_j + e_{jk}$$

Test the hypothesis that there is no change in weight; that is, test

$$H_0 : \mu_1 = \mu_2$$

against

$$H_1 : \mu_1 \text{ and } \mu_2 \text{ are not necessarily equal}$$

(*Hint:* this is the same as comparing Models 1 and 0 (equations 2.1 and 2.3) in the plant growth example.)

(b) Let $D_k = Y_{1k} - Y_{2k}$ for $k = 1, \ldots, K$. Show that this is of the form

Model 1: $$D_k = \mu + e_k \qquad (2.7)$$

and hence that another test of H_0 against H_1 is obtained by comparing this model with Model 0:

$$D_k = e_k \tag{2.8}$$

Assume that the random variables e_k are independent and all have the same distribution $N(0, \sigma^2)$ and hence find the maximum likelihood estimate of μ for Model 1 (equation 2.7).

Test H_0 against H_1 by comparing the values of S terms from these models.

(c) The analysis in (a) above is a two-sample (or unpaired) test. The analysis in (b) is a paired test which makes use of the natural relationship between weights of the *same* person before and after a diet. Are the conclusions the same for both analyses or different?

(d) List all the assumptions you made for the analyses in (a) and (b). How do they differ for the two analyses? Which analysis was more appropriate?

2.5 Suppose you have the following data:

x: 1.0	1.2	1.4	1.6	1.8	2.0
y: 3.15	4.85	6.50	7.20	8.25	13.50

and you want to fit the model

$$Y = \beta_0 + \beta_1 x + \beta_2 x^2 + e$$

If the model is expressed in matrix notation $\mathbf{y} = \boldsymbol{X\beta} + \mathbf{e}$ write down the vector and matrix terms $\mathbf{y}$, $\boldsymbol{X}$, $\boldsymbol{\beta}$ and $\mathbf{e}$.

2.6 Write in the notation $\mathbf{y} = \boldsymbol{X\beta} + \mathbf{e}$ the two-factor analysis of variance model

$$Y_{jk} = \mu + \alpha_j + \beta_k + e_{jk}$$

where $j = 1, 2$, $k = 1, 2, 3$, $\alpha_1 + \alpha_2 = 0$ and $\beta_1 + \beta_2 + \beta_3 = 0$ using the parameters μ, α_1, β_1 and β_2. (*Hint:* $\alpha_2 = -\alpha_1$ and $\beta_3 = -\beta_1 - \beta_2$.)

3

Exponential family of distributions and generalized linear models

3.1 INTRODUCTION

For several decades linear models of the form

$$\mathbf{y} = X\boldsymbol{\beta} + \mathbf{e} \tag{3.1}$$

in which the elements of $\mathbf{e}$ are assumed to be independent and identically distributed with the Normal distribution $N(0, \sigma^2)$, have formed the basis of most analyses of continuous data. For instance, in Chapter 2 the comparison of two means (plant growth example) and the relationship between a continuous response variable and a covariate in two groups (birthweight example) were analysed using models of this form. Generalizations of these examples to comparisons of more than two means (analysis of variance) and the relationship between a continuous response variable and several explanatory variables (multiple regression) are also of this form.

Recent advances in statistical theory and computer software allow us to use methods analogous to those developed for linear models in the following situations:

1. The response variables have distributions other than the Normal distribution – they may even be categorical rather than continuous;
2. The relationship between the response and explanatory variables need not be of the simple linear form in (3.1).

One of these advances has been the recognition that many of the 'nice' properties of the Normal distribution are shared by a wider class of distributions called the **exponential family of distributions**. These distributions and their properties are discussed in section 3.2.

A second advance is the extension of the numerical methods for estimating parameters, from linear combinations like $X\boldsymbol{\beta}$ in (3.1) to functions of linear combinations $g(X\boldsymbol{\beta})$. In theory the estimation procedures are straightforward. In practice they involve a considerable amount of computation so that they have only become feasible with the development of computer programs for numerical optimization of non-linear functions (Chambers, 1973). These are now included in many statistical packages. In particular, the program GLIM follows the same approach as this book and is referred to frequently. Details of the use of GLIM are given in *NAG, GLIM Manual* (1985), Healy (1988) and Aitkin *et al.* (1989).

This chapter introduces the exponential family of distributions and defines generalized linear models. Methods for parameter estimation and hypothesis testing are developed in Chapters 4 and 5 respectively. Some of the mathematical results are given in the appendices rather than the main text in order to maintain the continuity of the statistical development.

3.2 EXPONENTIAL FAMILY OF DISTRIBUTIONS

Consider a single random variable Y whose probability function, if it is discrete, or probability density function, if it is continuous, depends on a single parameter θ. The distribution belongs to the exponential family if it can be written in the form

$$f(y;\theta) = s(y)t(\theta)\mathrm{e}^{a(y)b(\theta)} \tag{3.2}$$

where a, b, s and t are known functions. Notice the symmetry between y and θ. This is emphasized if equation (3.2) is rewritten in the form

$$f(y;\theta) = \exp[a(y)b(\theta) + c(\theta) + d(y)] \tag{3.3}$$

where $s(y) = \exp d(y)$ and $t(\theta) = \exp c(\theta)$. For further details about these distributions see Barndorff-Nielsen (1978).

If $a(y) = y$, the distribution in (3.3) is said to be in the **canonical form** and $b(\theta)$ is sometimes called the **natural parameter** of the distribution.

If there are other parameters in addition to the parameter of interest θ they are regarded as **nuisance parameters** forming parts of the functions a, b, c and d, and they are treated as though they are known.

Many well-known distributions belong to the exponential family. For example, the Poisson, Normal and binomial distributions can all be written in the canonical form.

3.2.1 Poisson distribution

The probability function for the discrete random variable Y is

$$f(y;\lambda) = \frac{\lambda^y \mathrm{e}^{-\lambda}}{y!}$$

where y takes the values 0, 1, 2, ... This can be rewritten as

$$f(y; \lambda) = \exp[y \log \lambda - \lambda - \log y!]$$

which is in the canonical form with $\log \lambda$ as the natural parameter.

3.2.2 Normal distribution

The probability density function is

$$f(y; \mu) = \frac{1}{(2\pi\sigma^2)^{1/2}} \exp\left[-\frac{1}{2\sigma^2}(y - \mu)^2\right]$$

where μ is the parameter of interest and σ^2 is regarded as a nuisance parameter. This can be rewritten in the form

$$f(y; \mu) = \exp\left[-\frac{y^2}{2\sigma^2} + \frac{y\mu}{\sigma^2} - \frac{\mu^2}{2\sigma^2} - \frac{1}{2}\log(2\pi\sigma^2)\right]$$

This is in the canonical form. The natural parameter is $b(\mu) = \mu/\sigma^2$ and the other terms in (3.3) are

$$c(\mu) = -\frac{\mu^2}{2\sigma^2} - \frac{1}{2}\log(2\pi\sigma^2) \quad \text{and} \quad d(y) = -\frac{y^2}{2\sigma^2}$$

3.2.3 Binomial distribution

Let the random variable Y be the number of 'successes' in n independent trials in which the probability of success, π, is the same in all trials. Then Y has the binomial distribution with probability function

$$f(y; \pi) = \binom{n}{y} \pi^y (1 - \pi)^{n-y}$$

where y takes the values $0, 1, 2, \ldots, n$. This is denoted by $Y \sim b(n; \pi)$. Here π is the parameter of interest and n is assumed to be known. The probability function can be rewritten as

$$f(y; \pi) = \exp\left[y \log \pi - y \log(1 - \pi) + n \log(1 - \pi) + \log\binom{n}{y}\right]$$

which is of the form in equation (3.3).

These results are summarized in Table 3.1.

Other examples of distributions belonging to the exponential family are given in the exercises at the end of the chapter. Not all of them are of the canonical form.

We need to find expressions for the expected value and variance of $a(Y)$. To do this we use the following results which are derived in Appendix A. Let l be the log-likelihood function and U the first

Table 3.1 Poisson, Normal and binomial distributions as members of the exponential family

Distribution	Natural parameter	c	d
Poisson	$\log \lambda$	$-\lambda$	$-\log y!$
Normal	μ/σ^2	$-\frac{1}{2}\mu^2/\sigma^2 - \frac{1}{2}\log(2\pi\sigma^2)$	$-\frac{1}{2}y^2/\sigma^2$
Binomial	$\log\left(\dfrac{\pi}{1-\pi}\right)$	$n\log(1-\pi)$	$\log\begin{pmatrix} n \\ y \end{pmatrix}$

derivative of l with respect to θ, i.e. $U = \mathrm{d}l/\mathrm{d}\theta$. Then for any distribution the following results hold:

$$E(U) = 0 \tag{A.2}$$

and

$$\operatorname{var}(U) = E(U^2) = E(-U') \tag{A.3}$$

where the prime $'$ denotes the derivative with respect to θ. Here U is called the **score** and $\operatorname{var}(U)$ is called the **information**.

We can use these results for distributions in the exponential family. From (3.3) the log-likelihood function is

$$l = \log f = a(y)b(\theta) + c(\theta) + d(y)$$

so that

$$U = \frac{\mathrm{d}l}{\mathrm{d}\theta} = a(y)b'(\theta) + c'(\theta)$$

and

$$U' = \frac{\mathrm{d}^2 l}{\mathrm{d}\theta^2} = a(y)b''(\theta) + c''(\theta)$$

Thus

$$E(U) = b'(\theta)E[a(Y)] + c'(\theta)$$

but $E(U) = 0$, by result (A.2), so that

$$E[a(Y)] = -c'(\theta)/b'(\theta) \tag{3.4}$$

Also

$$\operatorname{var}(U) = [b'(\theta)]^2 \operatorname{var}[a(Y)]$$

and

$$E(-U') = -b''(\theta)E[a(Y)] - c''(\theta)$$

Now we use the result (A.3) to obtain

$$\begin{aligned}\text{var}[a(Y)] &= \{-b''(\theta)E[a(Y)] - c''(\theta)\}/[b'(\theta)]^2 \\ &= [b''(\theta)c'(\theta) - c''(\theta)b'(\theta)]/[b'(\theta)]^3 \qquad (3.5)\end{aligned}$$

It is easy to verify equations (3.4) and (3.5) for the Poisson, Normal and binomial distributions (see Exercise 3.4).

If $Y_1, \ldots, Y_N$ are independent random variables all with the same distribution given by (3.3), their joint probability density function is

$$\begin{aligned}f(y_1, \ldots, y_N) &= \prod_{i=1}^{N} \exp[b(\theta)a(y_i) + c(\theta) + d(y_i)] \\ &= \exp\left[b(\theta)\sum_{i=1}^{N} a(y_i) + Nc(\theta) + \sum_{i=1}^{N} d(y_i)\right]\end{aligned}$$

The term $\Sigma\, a(y_i)$ is said to be a **sufficient** statistic for $b(\theta)$; this means that in a certain sense $\Sigma\, a(y_i)$ summarizes all the available information about the parameter θ (Cox and Hinkley, 1974, Ch. 2). This is important for parameter estimation.

Next we consider a class of models based on the exponential family of distributions.

3.3 GENERALIZED LINEAR MODELS

The unity of many statistical methods involving linear combinations of parameters was demonstrated by Nelder and Wedderburn (1972) using the idea of a generalized linear model. This is defined in terms of a set of independent random variables $Y_1, \ldots, Y_N$ each with a distribution from the exponential family with the following properties:

1. The distribution of each Y_i is of the canonical form and depends on a single parameter θ_i (the θ_i's do not all have to be the same), thus

$$f(y_i;\, \theta_i) = \exp[y_i b_i(\theta_i) + c_i(\theta_i) + d_i(y_i)]$$

2. The distributions of all the Y_i's are of the same form (e.g. all Normal or all binomial) so that the subscripts on b, c and d are not needed.

Thus the joint probability density function of $Y_1, \ldots, Y_N$ is

$$f(y_1, \ldots, y_N;\, \theta_1, \ldots, \theta_N) = \exp\left[\sum_{i=1}^{N} y_i b(\theta_i) + \sum_{i=1}^{N} c(\theta_i) + \sum_{i=1}^{N} d(y_i)\right] \qquad (3.6)$$

For model specification, the parameters θ_i are usually not of direct interest (since there may be one for each observation). For a generalized linear model we consider a smaller set of parameters $\beta_1, \ldots, \beta_p$ (where $p < N$) such that a linear combination of the β's is equal to some function of the expected value μ_i of Y_i, i.e.

$$g(\mu_i) = \mathbf{x}_i^T \boldsymbol{\beta}$$

where g is a monotone, differentiable function called the **link function**;
$\mathbf{x}_i$ is a $p \times 1$ vector of explanatory variables (covariates and dummy variables for levels of factors);
$\boldsymbol{\beta}$ is the $p \times 1$ vector of parameters

$$\boldsymbol{\beta} = \begin{bmatrix} \beta_1 \\ \vdots \\ \beta_p \end{bmatrix}$$

Thus a generalized linear model has three components:

1. Response variables $Y_1, \ldots, Y_N$ which are assumed to share the same distribution from the exponential family;
2. A set of parameters $\boldsymbol{\beta}$ and explanatory variables

$$X = \begin{bmatrix} \mathbf{x}_1^T \\ \vdots \\ \mathbf{x}_N^T \end{bmatrix}$$

3. A monotone link function g such that

$$g(\mu_i) = \mathbf{x}_i^T \boldsymbol{\beta}$$

where

$$\mu_i = E(Y_i)$$

This chapter concludes with three examples of generalized linear models.

3.4 EXAMPLES

Example 3.1 Linear model
One special case is the linear model

$$\mathbf{y} = X\boldsymbol{\beta} + \mathbf{e}$$

where the elements e_i of $\mathbf{e}$ are independent and all have the distribution $N(0, \sigma^2)$. This is a generalized linear model because the elements of $\mathbf{y}$ are independent random variables Y_i with distributions $N(\mu_i, \sigma^2)$ where $\mu_i = \mathbf{x}_i^T \boldsymbol{\beta}$, the Normal distribution is a member of the exponential family (provided σ^2 is regarded as known) and, in this case, g is the identity function, i.e. $g(\mu_i) = \mu_i$. All the models discussed in Chapter 2 are of this form.

Example 3.2 Historical linguistics
Consider a language which is the descendant of another language as, for example, modern Greek is a descendant of ancient Greek, or the Romance languages are descendants of Latin. A simple model for the change in vocabulary is that if the languages are separated by time t then the probability that they have cognate words for a particular meaning is $e^{-\theta t}$ where θ is a parameter. It is claimed that θ is approximately the same for many commonly used meanings. For a test list of N different commonly used meanings suppose that a linguist judges, for each meaning, whether the corresponding words in two languages are cognate or not cognate. We can develop a generalized linear model to describe this situation.

Define random variables $Y_1, \ldots, Y_N$ as follows:

$$Y_i = \begin{cases} 1 & \text{if the languages have cognate words for meaning } i \\ 0 & \text{if the words are not cognate} \end{cases}$$

Then

$$P(Y_i = 1) = e^{-\theta t}$$

and

$$P(Y_i = 0) = 1 - e^{-\theta t}$$

This is a special case of the binomial distribution $b(n, \pi)$ with $n = 1$ and $E(Y_i) = \pi = e^{-\theta t}$.

The link function g in general is defined so that if $E(Y) = \mu$ then $g(\mu)$ is a linear combination of the parameters of interest. So in this case g is taken as the logarithmic function so that

$$g(\pi) = \log \pi = -\theta t.$$

Thus in the notation used above $\mathbf{x}_i = [-t]$ (the same for all i) and $\boldsymbol{\beta} = [\theta]$.

Example 3.3 Mortality trends
For a large population the probability of a randomly chosen individual dying from a given disease at a particular time is small. If we assume that the deaths among different individuals are independent events then the number of deaths, Y, in a fixed time period can be modelled by a Poisson distribution

$$f(y; \lambda) = \frac{\lambda^y e^{-\lambda}}{y!}$$

where y can take the values $0, 1, 2, \ldots$ and λ is the mean number of deaths per time period.

Trends in mortality can be modelled by taking independent random variables $Y_1, \ldots, Y_N$ to be the numbers of deaths occurring in successive time intervals numbered $i = 1, \ldots, N$. Let $E(Y_i) = \lambda_i$; typically this will vary with i.

The numbers of deaths from AIDS (acquired immunodeficiency syndrome) in Australia for three-month periods from 1983 to 1986 are shown in Table 3.2 and Fig. 3.1 (Whyte *et al.*, 1987).

Table 3.2 Numbers of deaths from AIDS in Australia per quarter in 1983–86 (the times $i = 1, \ldots, 14$ represent each of the three-month periods from January to March 1983 to April to June 1986, respectively)

i	1	2	3	4	5	6	7
y_i	0	1	2	3	1	4	9
i	8	9	10	11	12	13	14
y_i	18	23	31	20	25	37	45

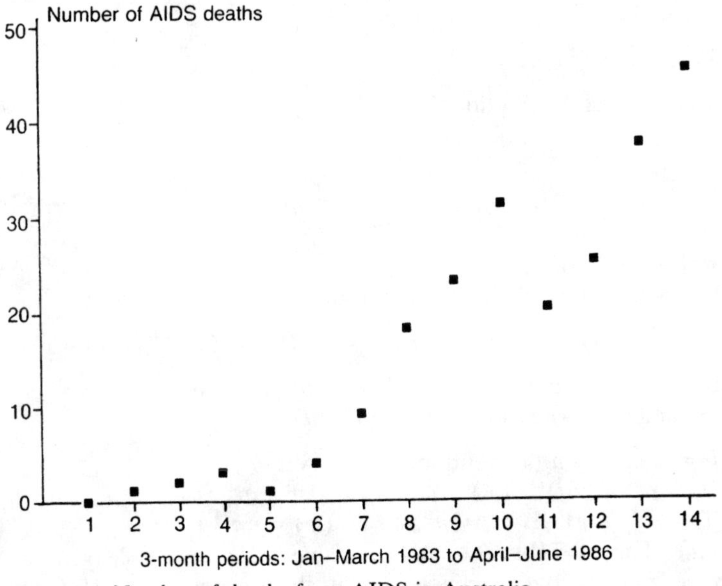

Figure 3.1 Number of deaths from AIDS in Australia.

Clearly the number of deaths is increasing with i. For these data a possible model is the Poisson distribution with

$$\lambda_i = i^\theta$$

where θ is a parameter to be estimated. This can be described by a generalized linear model in which the link function is

$$g(\lambda_i) = \log \lambda_i = \theta \log i$$

so $\mathbf{x}_i = [\log i]$ and $\boldsymbol{\beta} = [\theta]$.

3.5 EXERCISES

3.1 If the random variable Y has the gamma distribution with a scale parameter θ which is the parameter of interest and a known shape parameter ϕ, then its probability density function is

$$f(y;\theta) = \frac{y^{\phi-1}\theta^{\phi}e^{-y\theta}}{\Gamma(\phi)}$$

Show that this distribution belongs to the exponential family and hence find the natural parameter. Also using results in this chapter find $E(Y)$ and $\text{var}(Y)$.

3.2 Show that the following probability distributions belong to the exponential family:

(a) Pareto distribution $f(y;\theta) = \theta y^{-\theta-1}$

(b) Exponential distribution $f(y;\theta) = \theta e^{-y\theta}$

(c) Negative binomial distribution

$$f(y;\theta) = \binom{y+r-1}{r-1}\theta^r(1-\theta)^y$$

where r is known.

3.3 For the binomial distribution show from first principles that

$$E(U) = 0 \quad \text{and} \quad \text{var}(U) = E(U^2) = E(-U')$$

where $U = dl/d\theta$ and l is the log-likelihood function.

3.4 Use equations (3.4) and (3.5) to verify these results:

(a) For the Poisson distribution, $E(Y) = \text{var}(Y)$.
(b) If $Y \sim N(\mu, \sigma^2)$, $E(Y) = \mu$ and $\text{var}(Y) = \sigma^2$.
(c) If $Y \sim b(n, \pi)$, $E(Y) = n\pi$ and $\text{var}(Y) = n\pi(1-\pi)$.

3.5 Plot the data in Table 3.2 on a suitable scale to enable you to estimate roughly the parameter θ in the proposed model with $\lambda_i = i^{\theta}$. Use this value to estimate the expected values $E(Y_i)$ for each i and compare the estimates with the observed values y_i. Does this model appear to fit the data?

3.6 Consider N independent binary random variables $Y_1, \ldots, Y_N$ such that

$$P(Y_i = 1) = \pi_i \quad \text{and} \quad P(Y_i = 0) = 1 - \pi_i$$

The probability function of Y_i can be written as

$$\pi_i^{y_i}(1 - \pi_i)^{1-y_i}$$

where $y_i = 0$ or 1.

(a) Show that this probability function belongs to the exponential family of distributions.

(b) Show that the natural parameter is

$$\log\left(\frac{\pi_i}{1 - \pi_i}\right)$$

This function, the logarithm of the odds ratio $\pi_i/(1 - \pi_i)$, is called the **logit** function.

(c) Show that $E(Y_i) = \pi_i$.

(d) If the link function is defined as

$$g(\pi) = \log\left(\frac{\pi}{1 - \pi}\right) = \mathbf{x}^{\mathrm{T}}\boldsymbol{\beta}$$

show that this is equivalent to modelling the probability π as

$$\pi = \frac{\exp(\mathbf{x}^{\mathrm{T}}\boldsymbol{\beta})}{1 + \exp(\mathbf{x}^{\mathrm{T}}\boldsymbol{\beta})}$$

(e) In the particular case where $\mathbf{x}^{\mathrm{T}}\boldsymbol{\beta} = \beta_1 + \beta_2 x$ this gives

$$\pi = \frac{\exp(\beta_1 + \beta_2 x)}{1 + \exp(\beta_1 + \beta_2 x)}$$

which is the logistic function.

Sketch the graph of π against x in this case, taking β_1 and β_2 as constants. How would you interpret this if x is the dose of an insecticide and π is the probability of an insect dying?

3.7 Is the extreme value (Gumbel) distribution

$$f(y; \theta) = \frac{1}{\phi}\exp\left\{\left(\frac{y - \theta}{\phi}\right) - \exp[(y - \theta)/\phi]\right\}$$

(with $\phi > 0$ regarded as a nuisance parameter) a member of the exponential family?

3.8 Let $Y_1, \ldots, Y_N$ be independent random variables with

$$Y_i = \beta_0 + \log(\beta_1 + \beta_2 x_i) + e_i$$

where $e_i \sim N(0, \sigma^2)$ for all i. Is this a generalized linear model? Give reasons for your answer.

4

Estimation

4.1 INTRODUCTION

Two of the most commonly used approaches to the statistical estimation of parameters are the **method of maximum likelihood** and the **method of least squares**. This chapter begins by reviewing the principle of each of these methods and some properties of the estimators. Then the method of maximum likelihood is used for generalized linear models. Usually the estimates have to be obtained numerically by an iterative procedure which turns out to be closely related to weighted least squares estimation.

In Chapter 5 we consider the distributional properties of estimators for generalized linear models, including the calculation of standard errors and confidence regions, and also hypothesis testing.

4.2 METHOD OF MAXIMUM LIKELIHOOD

Let $Y_1, \ldots, Y_N$ be N random variables with the joint probability density function

$$f(y_1, \ldots, y_N; \theta_1, \ldots, \theta_p)$$

which depends on parameters $\theta_1, \ldots, \theta_p$. For brevity we denote

$$\begin{bmatrix} y_1 \\ \vdots \\ y_N \end{bmatrix} \text{by } \mathbf{y} \quad \text{and} \quad \begin{bmatrix} \theta_1 \\ \vdots \\ \theta_p \end{bmatrix} \text{by } \boldsymbol{\theta}$$

so the probability density function is denoted by $f(\mathbf{y}; \boldsymbol{\theta})$.

The **likelihood function** $L(\boldsymbol{\theta}; \mathbf{y})$ is algebraically the same as $f(\mathbf{y}; \boldsymbol{\theta})$ but the change in notation reflects a shift of emphasis from the random variables $\mathbf{y}$, with $\boldsymbol{\theta}$ fixed, to the parameters $\boldsymbol{\theta}$ with $\mathbf{y}$ fixed (where $\mathbf{y}$ represents the observations). Let Ω denote the set of all possible values of the parameter vector $\boldsymbol{\theta}$ (Ω is called the parameter space). The **maximum likelihood estimator** of $\boldsymbol{\theta}$ is the value $\hat{\boldsymbol{\theta}}$ which maximizes the likelihood function, that is

$$L(\hat{\boldsymbol{\theta}}; \mathbf{y}) \geqslant L(\boldsymbol{\theta}; \mathbf{y}) \quad \text{for all } \boldsymbol{\theta} \text{ in } \Omega$$

Equivalently, $\hat{\boldsymbol{\theta}}$ is the value which maximizes the log-likelihood function $l(\boldsymbol{\theta}; \mathbf{y}) = \log L(\boldsymbol{\theta}; \mathbf{y})$ (since the logarithmic function is monotonic). Thus

$$l(\hat{\boldsymbol{\theta}}; \mathbf{y}) \geqslant l(\boldsymbol{\theta}; \mathbf{y}) \quad \text{for all } \boldsymbol{\theta} \text{ in } \Omega$$

Often it is easier to work with the log-likelihood function than with the likelihood function itself.

Usually the estimator $\hat{\boldsymbol{\theta}}$ is obtained by differentiating the log-likelihood function with respect to each element θ_j of $\boldsymbol{\theta}$ and solving the simultaneous equations

$$\frac{\partial l(\boldsymbol{\theta}; \mathbf{y})}{\partial \theta_j} = 0 \qquad \text{for } j = 1, \ldots, p$$

It is necessary to check that the solutions do correspond to maxima of $l(\boldsymbol{\theta}; \mathbf{y})$ by verifying that the matrix of second derivatives

$$\frac{\partial^2 l(\boldsymbol{\theta}; \mathbf{y})}{\partial \theta_j \partial \theta_k}$$

evaluated at $\boldsymbol{\theta} = \hat{\boldsymbol{\theta}}$ is negative definite (e.g. if there is only one parameter θ to check that

$$\frac{\partial^2 l(\theta; \mathbf{y})}{\partial \theta^2}$$

evaluated at $\theta = \hat{\theta}$ is negative).

Also it is necessary to check if there are any values of $\boldsymbol{\theta}$ at the edges of the parameter space Ω which give local maxima of $l(\boldsymbol{\theta}; \mathbf{y})$. When all local maxima have been identified, the value of $\hat{\boldsymbol{\theta}}$ corresponding to the largest one is the maximum likelihood estimator. (For most of the models considered in this book there is only one maximum and it corresponds to the solution of the equations $\partial l / \partial \theta_j = 0$, $j = 1, \ldots, p$.)

An important property of maximum likelihood estimators is that if $g(\boldsymbol{\theta})$ is any function of the parameters $\boldsymbol{\theta}$, then the maximum likelihood estimator of $g(\boldsymbol{\theta})$ is $g(\hat{\boldsymbol{\theta}})$. This follows from the definition of $\hat{\boldsymbol{\theta}}$. It is sometimes called the **invariance property** of maximum likelihood estimators. A consequence is that we can work with any function of the parameters which is convenient for maximum likelihood estimation and then use the invariance property to obtain maximum likelihood estimates for the required parameters.

Other properties of maximum likelihood estimators include consistency, sufficiency and asymptotic efficiency. These are discussed in detail in books on theoretical statistics, for example Cox and Hinkley (1974, Ch. 9).

4.3 METHOD OF LEAST SQUARES

Let $Y_1, \ldots, Y_N$ be random variables with expected values

$$E(Y_i) = \mu_i \qquad \text{for } i = 1, \ldots, N$$

and let the μ_i's be functions of the parameters $\beta_1, \ldots, \beta_p$ (where $p < N$) which are to be estimated. Let

$$\boldsymbol{\beta} = \begin{bmatrix} \beta_1 \\ \vdots \\ \beta_p \end{bmatrix}$$

Consider the formulation

$$Y_i = \mu_i + e_i \qquad \text{for } i = 1, \ldots, N$$

in which μ_i represents the 'signal' component of Y_i and e_i represents the 'noise' component. The method of least squares consists of finding estimators $\hat{\boldsymbol{\beta}}$, also denoted by **b**, which minimize the sum of squares of the error terms e_i; that is, it involves minimizing the function

$$S = \sum e_i^2 = \sum [Y_i - \mu_i(\boldsymbol{\beta})]^2 \tag{4.1}$$

In matrix notation this is

$$S = (\mathbf{y} - \boldsymbol{\mu})^{\mathrm{T}}(\mathbf{y} - \boldsymbol{\mu})$$

where

$$\mathbf{y} = \begin{bmatrix} Y_1 \\ \vdots \\ Y_N \end{bmatrix} \quad \text{and} \quad \boldsymbol{\mu} = \begin{bmatrix} \mu_1 \\ \vdots \\ \mu_N \end{bmatrix}$$

Usually the estimator $\hat{\boldsymbol{\beta}}$ is obtained by differentiating S with respect to each element β_j of $\boldsymbol{\beta}$ and solving the simultaneous equation

$$\frac{\partial S}{\partial \beta_j} = 0 \qquad j = 1, \ldots, p$$

Of course it is necessary to check that the solutions correspond to minima (i.e. the matrix of second derivatives is positive definite) and to identify the global minimum from among these solutions and any local minima at the boundaries of the parameter space.

In practice there may be additional information about the Y_i's, for example that some observations are less reliable (i.e. have larger variance) than others. In such a case we may wish to weight the terms in (4.1) accordingly and minimize the sum

$$S_w = \sum w_i[Y_i - \mu_i(\boldsymbol{\beta})]^2$$

where the terms w_i represent weights, e.g. $w_i = [\text{var}(Y_i)]^{-1}$

More generally the Y_i's may be correlated; let V denote their variance–covariance matrix. Then **weighted least squares** estimators are obtained by minimizing

$$S_w = (\mathbf{y} - \boldsymbol{\mu})^{\mathrm{T}} V^{-1} (\mathbf{y} - \boldsymbol{\mu})$$

In particular if the terms μ_i are linear combinations of parameters β_j ($j = 1, \ldots, p$ where $p < N$) that is, if $\boldsymbol{\mu} = X\boldsymbol{\beta}$ for some $N \times p$ matrix X, then

$$S_w = (\mathbf{y} - X\boldsymbol{\beta})^{\mathrm{T}} V^{-1} (\mathbf{y} - X\boldsymbol{\beta}) \tag{4.2}$$

The derivatives of S_w with respect to the elements β_j of $\boldsymbol{\beta}$ are the vector

$$\frac{\partial S_w}{\partial \boldsymbol{\beta}} = -2X^{\mathrm{T}} V^{-1} (\mathbf{y} - X\boldsymbol{\beta})$$

so the weighted least squares estimator $\mathbf{b}$ of the parameter vector $\boldsymbol{\beta}$ is the solution of the **Normal equations**

$$X^{\mathrm{T}} V^{-1} X \mathbf{b} = X^{\mathrm{T}} V^{-1} \mathbf{y} \tag{4.3}$$

(as it can also be shown that the matrix of second derivatives is positive definite).

An important distinction between the methods of least squares and maximum likelihood is that least squares can be used without making assumptions about the distributions of the response variables Y_i beyond specifying their expectations and possibly their variance–covariance structure. In contrast, to obtain maximum likelihood estimators we need to specify the joint probability distribution of the Y_i's. However, to obtain the sampling distribution of the least squares estimators $\mathbf{b}$ additional assumptions about the Y_i's are generally required. Thus in practice there is little advantage in using the method of least squares unless the estimation equations are computationally simpler.

4.4 ESTIMATION FOR GENERALIZED LINEAR MODELS

We wish to obtain the maximum likelihood estimators of the parameters $\boldsymbol{\beta}$ for the generalized linear models defined in Section 3.3. The log-likelihood function for independent responses $Y_1, \ldots, Y_N$ is

$$l(\boldsymbol{\theta}; \mathbf{y}) = \sum y_i b(\theta_i) + \sum c(\theta_i) + \sum d(y_i)$$

where

$$E(Y_i) = \mu_i = -c'(\theta_i)/b'(\theta_i)$$

and

$$g(\mu_i) = \mathbf{x}_i^{\mathrm{T}}\boldsymbol{\beta} = \eta_i$$

where g is some monotone and differentiable function.

A property of the exponential family of distributions is that they satisfy enough regularity conditions to ensure that the global maximum of the log-likelihood function $l(\boldsymbol{\theta}; \mathbf{y})$ is given uniquely by the solution of the equations $\partial l/\partial \boldsymbol{\theta} = \mathbf{0}$, or equivalently by the solutions of $\partial l/\partial \boldsymbol{\beta} = \mathbf{0}$ (see Cox and Hinkley, 1974, Ch. 9).

In Appendix B (A.12) it is shown that

$$\frac{\partial l}{\partial \beta_j} = U_j = \sum_{i=1}^{N} \frac{(y_i - \mu_i)x_{ij}}{\operatorname{var}(Y_i)} \left(\frac{\partial \mu_i}{\partial \eta_i}\right) \tag{4.4}$$

where x_{ij} is the jth element of $\mathbf{x}_i^{\mathrm{T}}$. In general the equations $U_j = 0$ ($j = 1, \ldots, p$) are non-linear and they have to be solved by numerical iteration. If the **Newton–Raphson method** is used then the mth approximation is given by

$$\mathbf{b}^{(m)} = \mathbf{b}^{(m-1)} - \left[\frac{\partial^2 l}{\partial \beta_j \partial \beta_k}\right]^{-1}_{\boldsymbol{\beta}=\mathbf{b}^{(m-1)}} \mathbf{U}^{(m-1)} \tag{4.5}$$

where

$$\left[\frac{\partial^2 l}{\partial \beta_j \partial \beta_k}\right]_{\boldsymbol{\beta}=\mathbf{b}^{(m-1)}}$$

is the matrix of second derivatives of l evaluated at $\boldsymbol{\beta} = \mathbf{b}^{(m-1)}$ and $\mathbf{U}^{(m-1)}$ is the vector of first derivatives $U_j = \partial l/\partial \beta_j$ evaluated at $\boldsymbol{\beta} = \mathbf{b}^{(m-1)}$. (This is the multidimensional version of the Newton–Raphson method for finding a solution of a single variable equation $f(x) = 0$, namely

$$x^{(m)} = x^{(m-1)} - \frac{f[x^{(m-1)}]}{f'[x^{(m-1)}]}\Bigg)$$

An alternative procedure which is sometimes simpler than the Newton–Raphson method is called the **method of scoring**. It involves replacing the matrix of second derivatives in (4.5) by the matrix of expected values

$$E\left[\frac{\partial^2 l}{\partial \beta_j \partial \beta_k}\right]$$

In Appendix A it is shown that this is equal to the negative of the variance–covariance matrix of the U_j's. The **information matrix** $\mathcal{J} = E[\mathbf{U}\mathbf{U}^{\mathrm{T}}]$ has the elements

$$\mathcal{J}_{jk} = E[U_j U_k] = E\left[\frac{\partial l}{\partial \beta_j} \frac{\partial l}{\partial \beta_k}\right]$$

$$= -E\left[\frac{\partial^2 l}{\partial\beta_j \partial\beta_k}\right]$$

by result (A.6). Thus equation (4.5) is replaced by

$$\mathbf{b}^{(m)} = \mathbf{b}^{(m-1)} + [\mathcal{J}^{(m-1)}]^{-1}\mathbf{U}^{(m-1)} \tag{4.6}$$

where $\mathcal{J}^{(m-1)}$ denotes the information matrix evaluated at $\mathbf{b}^{(m-1)}$. If both sides of equation (4.6) are multiplied by $\mathcal{J}^{(m-1)}$ we obtain

$$\mathcal{J}^{(m-1)}\mathbf{b}^{(m)} = \mathcal{J}^{(m-1)}\mathbf{b}^{(m-1)} + \mathbf{U}^{(m-1)} \tag{4.7}$$

For generalized linear models the (j, k)th element of $\mathcal{J}$ is

$$\mathcal{J}_{jk} = \sum_{i=1}^{N} \frac{x_{ij}x_{ik}}{\text{var}(Y_i)}\left(\frac{\partial\mu_i}{\partial\eta_i}\right)^2 \tag{4.8}$$

(see Appendix B (A.13)). Thus $\mathcal{J}$ can be written as

$$\mathcal{J} = \boldsymbol{X}^{\text{T}}\boldsymbol{W}\boldsymbol{X}$$

where $\boldsymbol{W}$ is the $N \times N$ diagonal matrix with elements

$$w_{ii} = \frac{1}{\text{var}(Y_i)}\left(\frac{\partial\mu_i}{\partial\eta_i}\right)^2 \tag{4.9}$$

The expression on the right-hand side of (4.7) is the vector with elements

$$\sum_k \sum_i \frac{x_{ij}x_{ik}}{\text{var}(Y_i)}\left(\frac{\partial\mu_i}{\partial\eta_i}\right)^2 b_k^{(m-1)} + \sum_i \frac{(y_i - \mu_i)x_{ij}}{\text{var}(Y_i)}\left(\frac{\partial\mu_i}{\partial\eta_i}\right)$$

evaluated at $\mathbf{b}^{(m-1)}$; this follows from equations (4.8) and (4.4). Thus the right-hand side of equation (4.7) can be written as

$$\boldsymbol{X}^{\text{T}}\boldsymbol{W}\mathbf{z}$$

where $\mathbf{z}$ has elements

$$z_i = \sum_k x_{ik}b_k^{(m-1)} + (y_i - \mu_i)\left(\frac{\partial\eta_i}{\partial\mu_i}\right) \tag{4.10}$$

with μ_i and $\partial\eta_i/\partial\mu_i$ evaluated at $\mathbf{b}^{(m-1)}$.

Hence the iterative equation for the method of scoring, (4.7), can be written as

$$\boldsymbol{X}^{\text{T}}\boldsymbol{W}\boldsymbol{X}\mathbf{b}^{(m)} = \boldsymbol{X}^{\text{T}}\boldsymbol{W}\mathbf{z} \tag{4.11}$$

This has the same form as the normal equations for a linear model obtained by weighted least squares, (4.3), except that (4.11) has to be solved iteratively because in general $\mathbf{z}$ and $\boldsymbol{W}$ depend on $\mathbf{b}$. Thus for generalized linear models maximum likelihood estimators are obtained by an **iterative weighted least squares** procedure.

Usually a computer is needed to solve (4.11). Most statistical packages which include analyses based on generalized linear models have efficient programs for calculating the solutions. They begin by using some initial approximation $\mathbf{b}^{(0)}$ to evaluate $\mathbf{z}$ and W, then (4.11) is solved to give $\mathbf{b}^{(1)}$ which in turn is used to obtain better approximations for $\mathbf{z}$ and W, and so on until adequate convergence is achieved. When the difference between successive approximations $\mathbf{b}^{(m)}$ and $\mathbf{b}^{(m-1)}$ is sufficiently small, $\mathbf{b}^{(m)}$ is taken as the maximum likelihood estimate. The example below illustrates the use of this estimation procedure.

4.5 EXAMPLE OF SIMPLE LINEAR REGRESSION FOR POISSON RESPONSES

The data in Table 4.1 are counts y_i observed at various values of a covariate x. They are plotted in Fig. 4.1.

Table 4.1 Poisson regression data

y_i	2	3	6	7	8	9	10	12	15
x_i	−1	−1	0	0	0	0	1	1	1

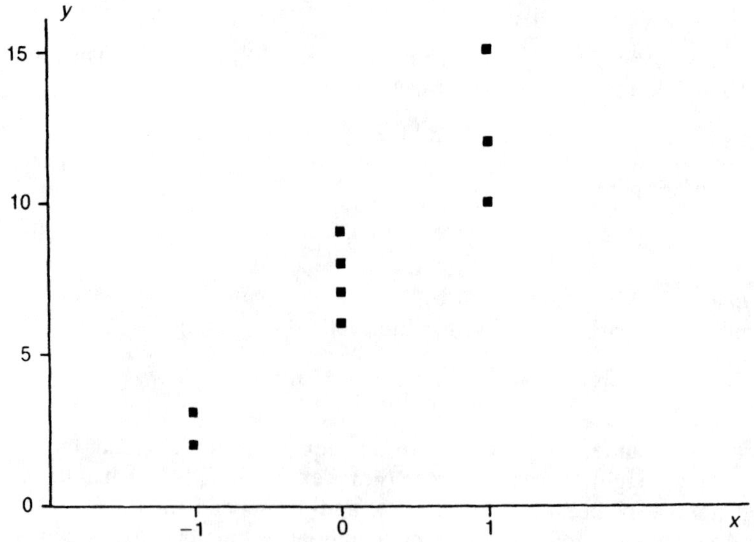

Figure 4.1 Plot of data in Table 4.1.

Let us assume that the responses Y_i are Poisson random variables. In practice, such an assumption would be made either on substantive grounds or from observing that the variability increases with Y. For the Poisson distribution, the expected value and variance of Y_i are equal

$$E(Y_i) = \text{var}(Y_i)$$

Let us model the relationship between Y_i and x_i by the straight line

$$\begin{aligned} E(Y_i) = \mu_i &= \beta_1 + \beta_2 x_i \\ &= \mathbf{x}_i^{\mathrm{T}}\boldsymbol{\beta} \end{aligned}$$

where

$$\boldsymbol{\beta} = \begin{bmatrix} \beta_1 \\ \beta_2 \end{bmatrix} \quad \text{and} \quad \mathbf{x}_i = \begin{bmatrix} 1 \\ x_i \end{bmatrix}$$

for $i = 1, \ldots, N$ where $N = 9$. Thus we take the link function $g(\mu_i)$ to be the identity function

$$g(\mu_i) = \mu_i = \mathbf{x}_i^{\mathrm{T}}\boldsymbol{\beta} = \eta_i$$

Therefore $\partial\mu_i/\partial\eta_i = 1$ which simplifies equations (4.9) and (4.10). From (4.9)

$$w_{ii} = \frac{1}{\text{var}(Y_i)} = \frac{1}{\beta_1 + \beta_2 x_i}$$

and from (4.10)

$$z_i = b_1 + b_2 x_i + (y_i - b_1 - b_2 x_i) = y_i$$

Hence using the estimate $\mathbf{b}$ for $\boldsymbol{\beta}$

$$\mathcal{J} = \boldsymbol{X}^{\mathrm{T}}\boldsymbol{W}\boldsymbol{X} = \begin{bmatrix} \sum_{i=1}^{N} \frac{1}{b_1 + b_2 x_i} & \sum_{i=1}^{N} \frac{x_i}{b_1 + b_2 x_i} \\ \sum_{i=1}^{N} \frac{x_i}{b_1 + b_2 x_i} & \sum_{i=1}^{N} \frac{x_i^2}{b_1 + b_2 x_i} \end{bmatrix}$$

and

$$\boldsymbol{X}^{\mathrm{T}}\boldsymbol{W}\mathbf{z} = \begin{bmatrix} \sum_{i=1}^{N} \frac{y_i}{b_1 + b_2 x_i} \\ \sum_{i=1}^{N} \frac{x_i y_i}{b_1 + b_2 x_i} \end{bmatrix}$$

The maximum likelihood estimates are obtained iteratively from the equations

$$(\boldsymbol{X}^{\mathrm{T}}\boldsymbol{W}\boldsymbol{X})^{(m-1)}\mathbf{b}^{(m)} = (\boldsymbol{X}^{\mathrm{T}}\boldsymbol{W}\mathbf{z})^{(m-1)}$$

where the superscript $(m-1)$ denotes evaluation at $\mathbf{b}^{(m-1)}$. From Fig. 4.1 we can choose initial values $b_1^{(0)} = 7$ and $b_2^{(0)} = 5$. Successive approximations are shown in Table 4.2. Thus the maximum likelihood estimates, correct to four decimal places, are $b_1 = 7.4516$ and $b_2 = 4.9353$.

Table 4.2 Successive approximations for regression coefficients

m	0	1	2	3
$b_1^{(m)}$	7	7.450	7.4516	7.4516
$b_2^{(m)}$	5	4.937	4.9353	4.9353

MINITAB instructions (version 5.1.3)	Comments
MTB > store 'poilinreg'	Name for the macro
STOR> multiply m2 m3 m4	M4 $\equiv \boldsymbol{X}\mathbf{b}$
STOR> copy m4 c1	C1 has elements $[b_1 + b_2x_i]$
STOR> let c2 = 1/c1	C2 has elements $[1/(b_1 + b_2x_i)]$
STOR> diagonal c2 m4	M4 $\equiv \boldsymbol{W}$
STOR> transpose m2 m5	M5 $\equiv \boldsymbol{X}^{\mathrm{T}}$
STOR> multiply m5 m4 m6	M6 $\equiv \boldsymbol{X}^{\mathrm{T}}\boldsymbol{W}$
STOR> multiply m6 m1 m7	M7 $\equiv \boldsymbol{X}^{\mathrm{T}}\boldsymbol{W}\mathbf{z}$
STOR> multiply m6 m2 m8	M8 $\equiv \boldsymbol{X}^{\mathrm{T}}\boldsymbol{W}\boldsymbol{X}$
STOR> invert m8 m9	M9 $\equiv (\boldsymbol{X}^{\mathrm{T}}\boldsymbol{W}\boldsymbol{X})^{-1}$
STOR> multiply m9 m7 m3	M3 $\equiv (\boldsymbol{X}^{\mathrm{T}}\boldsymbol{W}\boldsymbol{X})^{-1}(\boldsymbol{X}^{\mathrm{T}}\boldsymbol{W}\mathbf{z}) = \mathbf{b}$
STOR> end	

4.6 MINITAB PROGRAM FOR SIMPLE LINEAR REGRESSION WITH POISSON RESPONSES (adapted from Fox, 1986)

Estimation of parameters for generalized linear models requires iteration involving matrix transposition and multiplication, and the solution of sets of linear equations. The statistical program MINITAB can readily perform these operations; for further details of MINITAB, see Ryan, Joiner and Ryan (1985). The procedure is illustrated for the example in section 4.5 for the Poisson response variables Y_i with

$$E(Y_i) = \beta_1 + \beta_2 x_i$$

Suppose that the observations $\mathbf{y}$ have been read into the $N \times 1$ matrix M1, the covariate values x_i have been read into the second column of

the $N \times 2$ matrix M2 whose first column has been set to 1s, and the initial estimates for $\boldsymbol{\beta}$ have been read into the 2×1 matrix M3. The following 'macro' (i.e. stored set of instructions) will perform the iterative step.
For the data in the example in section 4.5 and the initial estimates $b_1^{(0)} = 7$ and $b_2^{(0)} = 5$ the matrices are

$$\text{M1} = \begin{bmatrix} 2 \\ 3 \\ \vdots \\ 15 \end{bmatrix} \quad \text{M2} = \begin{bmatrix} 1 & -1 \\ 1 & -1 \\ \vdots & \vdots \\ 1 & 1 \end{bmatrix} \quad \text{and} \quad \text{M3} = \begin{bmatrix} 7 \\ 5 \end{bmatrix}$$

The following MINITAB instructions will perform the iterative step five times and print the results

```
MTB > execute 'poilinreg' 5

MTB > print m3
```

The results obtained are $b_1 = 7.451\,63$ and $b_2 = 4.935\,30$ corresponding to the results shown to Table 4.2.

4.7 GLIM

Although the calculations for estimating parameters for particular generalized linear models can be readily programmed, as illustrated in section 4.6, programs of greater generality are required. These should allow for various response distributions (e.g. binomial, Poisson or Normal) and different link functions (e.g. logs, logits, etc.). They should allow the *design matrix* $\mathbf{X}$ to be specified easily. In addition, the programs should be accurate and efficient to use. (In contrast, matrix inversion to solve linear equations, as used in the MINITAB program above, is inefficient and potentially inaccurate.) The program GLIM meets all these requirements and relates closely to the approach developed in this book. Other programs for generalized linear modelling are available, especially in the major statistical computing systems (see section 1.1).

GLIM is an interactive program. First the numbers of observations, covariates, factor levels and so on, have to be set (in order to specify the sizes of matrices). Then the data are read in and elements of the design matrix $\mathbf{X}$ are set up. The next step is to choose the distribution and link function required. Once this is done the linear components, $\mathbf{x}^T\boldsymbol{\beta}$, of the model are 'fitted', that is, the parameter values $\boldsymbol{\beta}$ are estimated and the estimates, goodness-of-fit statistics (see Chapter 5) and other information can be displayed. More details about GLIM are given in *NAG, GLIM Manual* (1985), Healy (1988) and Aitkin *et al.* (1989).

For the example on simple linear regression with Poisson responses (section 4.5) the following GLIM input and output illustrate the estimation of the parameters.

GLIM (version 3.77)	Comments
? \$units 9\$	Length of vectors and matrices
? \$data y x \$	
? \$read	
\$REA? 2 −1	
\$REA? 3 −1	
⋮	
\$REA? 15 1	
? \$yvar y \$	y is the response variable
? \$error poisson \$	Specify the distribution
? \$link identity \$	Specify the link
? \$fit x \$	Fit the covariate x
scaled deviance = 1.8947 at cycle 3 d.f. = 7	The output – three iterations were needed
? \$display e\$	Display the estimates
estimate s.e. parameter	
1 7.452 0.8841 1	Estimate of β_1
2 4.935 1.089 X	Estimate of β_2
scale parameter taken as 1.000	

Thus the parameter estimates produced by GLIM agree with those obtained in sections 4.5 and 4.6.

4.8 EXERCISES

4.1 The data in Table 4.3 (from Table 3.2) show the numbers of deaths from AIDS in Australia for successive three-month periods from 1983 to 1986.

Table 4.3 Numbers of deaths from AIDS in Australia per quarter from January–March 1983 to April–June 1986; y_i denotes the number of deaths and $x_i = \log i$ where $i = 1, \ldots, 14$ indicates the quarter

y_i	0	1	2	3	1	4	9
x_i	0	0.693	1.099	1.386	1.609	1.792	1.946
y_i	18	23	31	20	25	37	45
x_i	2.079	2.197	2.303	2.398	2.485	2.565	2.639

Suppose that the random variables Y_i are Poisson variables with $E(Y_i) = \mu_i$ where

$$g(\mu_i) = \log \mu_i = \beta_1 + \beta_2 x_i$$

and $x_i = \log i$ (this is slightly more general than the model proposed in Example 3.3 and Exercise 3.5). The link function used in this case is the logarithmic function (which is the 'natural' link for the Poisson distribution in the sense that it corresponds to the natural parameter, see Table 3.1).

(a) Use equations (4.9) and (4.10) to obtain expressions for the elements of $\mathbf{W}$ and $\mathbf{z}$ for this model.
(b) For the data in Table 4.3 estimate the parameters of the model by adapting the MINITAB macro given in section 4.6.
(c) Repeat the estimation using GLIM (see section 4.7). (In GLIM the natural link is the default setting, i.e. it need not be specified. Also you can use the commands CALC and %GL to generate the values of i and $x_i = \log i$.)

4.2 Let $Y_1, \ldots, Y_N$ be a random sample from the Normal distribution $N(\log \beta, \sigma^2)$ where σ^2 is known. Find the maximum likelihood estimator of β from first principles. Also verify equations (4.4) and (4.11) in this case.

4.3 The data in Table 4.4 are times to death, y_i, in weeks from diagnosis and $\log_{10}$ (initial white blood cell count), x_i, for seventeen patients suffering from leukaemia. (This is Example U from Cox and Snell, 1981).

Table 4.4 Survival time y_i in weeks and $\log_{10}$ (initial white blood cell count) x_i for seventeen leukaemia patients

y_i	65	156	100	134	16	108	121	4	39
x_i	3.36	2.88	3.63	3.41	3.78	4.02	4.00	4.23	3.73
y_i	143	56	26	22	1	1	5	65	
x_i	3.85	3.97	4.51	4.54	5.00	5.00	4.72	5.00	

(a) Plot y_i against x_i. Do the data show any trend?
(b) A possible specification for $E(Y)$ is

$$E(Y_i) = \exp(\beta_1 + \beta_2 x_i)$$

which will ensure that $E(Y)$ is non-negative for all values of the parameters and all values of x. Which link function is appropriate in this case?

(c) The exponential distribution is often used to describe survival times. Show that this is a special case of the gamma distribution (see Exercises 3.1 and 3.2(b)).

(d) Use GLIM to fit the model suggested in parts (b) and (c) above. (*Hint:* to model the exponential distribution rather than the more general gamma distribution it is necessary to specify that the shape parameter is 1. This can be done in GLIM using the commands ERROR GAMMA and SCALE 1.)

Plot the fitted model on the graph obtained in part (a).

Do you consider the model to be an adequate description of the data?

4.4 An alternative derivation of the Newton–Raphson equation (4.5) can be obtained by approximating the log-likelihood function $l(\boldsymbol{\beta}; \mathbf{y})$ by a Taylor series expansion about $\boldsymbol{\beta} = \boldsymbol{\beta}^*$. The equation used is

$$l(\boldsymbol{\beta}; \mathbf{y}) = l(\boldsymbol{\beta}^*; \mathbf{y}) + (\boldsymbol{\beta} - \boldsymbol{\beta}^*)^{\mathrm{T}}\mathbf{U} + (\boldsymbol{\beta} - \boldsymbol{\beta}^*)^{\mathrm{T}}\boldsymbol{H}(\boldsymbol{\beta} - \boldsymbol{\beta}^*)/2$$

where $\mathbf{U}$, the $p \times 1$ vector with elements $U_j = \partial l/\partial\beta_j$, and $\boldsymbol{H}$, the $p \times p$ matrix with elements $\partial^2 l/\partial\beta_j\partial\beta_k$, are evaluated at $\boldsymbol{\beta} = \boldsymbol{\beta}^*$.

(a) Write down the single parameter version of this approximation and use it to obtain an expression for the maximum likelihood estimator for β. If β^* is regarded as the $(m - 1)$th approximation and β as the mth approximation show that the equation corresponds to the single parameter version of (4.5).

(b) Prove the corresponding result for the general case.

4.5 Let $Y_1, \ldots, Y_N$ be independent random variables with $Y_i \sim N(\boldsymbol{x}_i^{\mathrm{T}}\boldsymbol{\beta}, \sigma_i^2)$. Show that the maximum likelihood estimator of $\boldsymbol{\beta}$ is the solution of $\boldsymbol{X}^{\mathrm{T}}\boldsymbol{V}^{-1}\boldsymbol{X}\mathbf{b} = \boldsymbol{X}^{\mathrm{T}}\boldsymbol{V}^{-1}\mathbf{y}$ where $\boldsymbol{V}$ is the diagonal matrix with elements $v_{ii} = \sigma_i^2$. (Since this is the same as equation (4.3), for linear models with normal errors, maximum likelihood estimators and least squares estimators are identical.)

5

Inference

5.1 INTRODUCTION

Statistical modelling involves three steps: (1) specifying models; (2) estimating parameters; (3) making inferences – that is, testing hypotheses, obtaining confidence intervals, and assessing the goodness of fit of models. Model specification for generalized linear models is discussed in Chapter 3 and parameter estimation in Chapter 4. This chapter covers the third step. It describes the sampling distributions of the estimators and of statistics for measuring goodness of fit and shows how these are used to make inferences.

In the particular case of linear models with Normally distributed response variables the sampling distributions can be determined exactly. In general, the problem of finding exact distributions is intractable and we rely instead on large-sample asymptotic results. The rigorous development of these results requires careful attention to various regularity conditions. For independent observations from distributions which belong to the exponential family, and in particular for generalized linear models, the necessary conditions are indeed satisfied. In this book we consider only the major steps and not the finer points involved in deriving the sampling distributions. For a further discussion of the principles, see, for example, Chapter 9 of Cox and Hinkley (1974). Theoretical results for generalized linear models are given by Fahrmeir and Kaufman (1985).

The basic idea is that if $\hat{\theta}$ is a consistent estimator of a parameter θ and $\text{var}(\hat{\theta})$ is the variance of the estimator then for large samples the following results hold, at least approximately:

1. $\hat{\theta}$ is an unbiased estimator of θ (because for a consistent estimator $E(\hat{\theta})$ approaches θ as the sample size becomes large);
2. The statistic

$$\frac{\hat{\theta} - \theta}{\sqrt{[\text{var}(\hat{\theta})]}}$$

has the standard Normal distribution $N(0, 1)$; or equivalently, the square of the statistic has a chi-squared distribution with one degree of freedom

$$\frac{(\hat{\theta} - \theta)^2}{\text{var}(\hat{\theta})} \sim \chi_1^2$$

The generalization of these results to p parameters is as follows. Let $\boldsymbol{\theta}$ be a vector of p parameters. Let $\hat{\boldsymbol{\theta}}$ be a consistent estimator of $\boldsymbol{\theta}$ and let $\boldsymbol{V}$ denote the variance–covariance matrix for $\hat{\boldsymbol{\theta}}$. Then asymptotically $\hat{\boldsymbol{\theta}}$ is an unbiased estimator of $\boldsymbol{\theta}$ and, provided the matrix $\boldsymbol{V}$ is non-singular so that its inverse exists, the sampling distribution is

$$(\hat{\boldsymbol{\theta}} - \boldsymbol{\theta})^{\mathrm{T}}\boldsymbol{V}^{-1}(\hat{\boldsymbol{\theta}} - \boldsymbol{\theta}) \sim \chi_p^2$$

by the definition of the central chi-squared distribution (1.2).

If the variance–covariance matrix is singular so that it does not have a unique inverse there are two approaches which can be used. Suppose that $\boldsymbol{V}$ has rank q where $q < p$. One approach is to obtain a generalized inverse $\boldsymbol{V}^{-}$, i.e. any matrix such that $\boldsymbol{V}\boldsymbol{V}^{-}\boldsymbol{V} = \boldsymbol{V}$, then it can be shown that asymptotically

$$(\hat{\boldsymbol{\theta}} - \boldsymbol{\theta})^{\mathrm{T}}\boldsymbol{V}^{-}(\hat{\boldsymbol{\theta}} - \boldsymbol{\theta}) \sim \chi_q^2$$

The other approach is to re-express the model in terms of a new parameter vector $\boldsymbol{\phi}$ of length q such that the variance–covariance matrix of $\boldsymbol{\phi}$, say $\boldsymbol{W}$, is non-singular then

$$(\hat{\boldsymbol{\phi}} - \boldsymbol{\phi})^{\mathrm{T}}\boldsymbol{W}^{-1}(\hat{\boldsymbol{\phi}} - \boldsymbol{\phi}) \sim \chi_q^2$$

Both of these approaches are used later in this book.

In this chapter we obtain the sampling distributions for the following statistics: the scores $U_j = \partial l/\partial \beta_j$, the maximum likelihood estimators b_j and a goodness-of-fit statistic derived from the likelihood ratio test. In each case the sampling distribution is used to make inferences about the fitted model. The final section of this chapter describes **residuals** which provide another useful way of examining how well the fitted model describes the actual data.

5.2 SAMPLING DISTRIBUTION FOR SCORES

The score statistic corresponding to a parameter β_j is defined as the derivative of the log-likelihood function l with respect to β_j so for a vector $\boldsymbol{\beta}$ of p parameters the scores are

$$U_j = \frac{\partial l}{\partial \beta_j} \qquad \text{for } j = 1, \ldots, p$$

For generalized linear models it is shown in Appendix A that

$$E(U_j) = 0$$

for all j and that the variance–covariance matrix of the U_j's is the information matrix $\mathcal{J}$, i.e.

$$E(\mathbf{U}\mathbf{U}^{\mathrm{T}}) = \mathcal{J}$$

where

$$\mathbf{U} = \begin{bmatrix} U_1 \\ \vdots \\ U_p \end{bmatrix}$$

Hence by the central limit theorem, at least asymptotically, $\mathbf{U}$ has the multivariate Normal distribution $\mathbf{U} \sim N(\mathbf{0}, \mathcal{J})$ and therefore

$$\mathbf{U}^{\mathrm{T}}\mathcal{J}^{-1}\mathbf{U} \sim \chi_p^2 \tag{5.1}$$

(provided that $\mathcal{J}$ is non-singular so that its inverse $\mathcal{J}^{-1}$ exists).

Example 5.1
This example shows that result (5.1) is exact for the Normal distribution. Let $Y_1, \ldots, Y_N$ denote N independent, identically distributed random variables with the distribution $N(\mu, \sigma^2)$ where σ^2 is a known constant. For the generalized linear model in this case, there is only one parameter of interest μ, there are no explanatory variables and the link function is the identity. The log-likelihood function is

$$l(\mu; y_1, \ldots, y_N) = -\frac{1}{2\sigma^2}\sum(y_i - \mu)^2 - N\log[\sigma\surd(2\pi)]$$

so that

$$U = \frac{\mathrm{d}l}{\mathrm{d}\mu} = \frac{1}{\sigma^2}\sum(y_i - \mu)$$

Thus the score statistic is

$$U = \frac{1}{\sigma^2}\sum(Y_i - \mu) = \frac{N}{\sigma^2}(\bar{Y} - \mu)$$

It is easy to see that $E(U) = 0$ because $E(\bar{Y}) = \mu$. The information $\mathcal{J}$ is given by

$$\mathcal{J} = \mathrm{var}(U) = \frac{N^2}{\sigma^4}\mathrm{var}(\bar{Y}) = \frac{N}{\sigma^2}$$

because $\mathrm{var}(\bar{Y}) = \sigma^2/N$. Therefore the statistic $U^{\mathrm{T}}\mathcal{J}^{-1}U$ is given by

$$U^{\mathrm{T}}\mathcal{J}^{-1}U = \left[\frac{N(\bar{Y} - \mu)}{\sigma^2}\right]^2 \frac{\sigma^2}{N}$$

$$= \frac{(\bar{Y} - \mu)^2}{\sigma^2/N}$$

But $\bar{Y} \sim N(\mu, \sigma^2/N)$ so $(\bar{Y} - \mu)^2/(\sigma^2/N) \sim \chi_1^2$. Therefore $U^T \mathcal{J}^{-1} U \sim \chi_1^2$ exactly. Either of the forms

$$\frac{(\bar{Y} - \mu)}{\sigma/\sqrt{N}} \sim N(0, 1) \quad \text{or} \quad \frac{(\bar{Y} - \mu)^2}{\sigma^2/N} \sim \chi_1^2$$

can be used to test hypotheses or obtain confidence intervals for μ.

Example 5.2

Let the response variable Y have the binomial distribution $b(n, \pi)$. The log-likelihood function is

$$l(y; \pi) = y \log \pi + (n - y) \log (1 - \pi) + \log \binom{n}{y}$$

so the score statistic obtained from $\mathrm{d}l/\mathrm{d}\pi$ is

$$U = \frac{Y}{\pi} - \frac{n - Y}{1 - \pi} = \frac{Y - n\pi}{\pi(1 - \pi)}$$

But $E(Y) = n\pi$ and so $E(U) = 0$. Also $\mathrm{var}(Y) = n\pi(1 - \pi)$ and so the information is

$$\mathcal{J} = \mathrm{var}(U) = \frac{1}{\pi^2(1 - \pi)^2} \mathrm{var}(Y) = \frac{n}{\pi(1 - \pi)}$$

Therefore

$$U^T \mathcal{J}^{-1} U = \frac{(Y - n\pi)^2}{\pi^2(1 - \pi)^2} \frac{\pi(1 - \pi)}{n}$$

$$= \frac{(Y - n\pi)^2}{n\pi(1 - \pi)}$$

Hence result (5.1) that $U^T \mathcal{J}^{-1} U \sim \chi_1^2$ is equivalent to the usual Normal approximation to the binomial distribution, i.e., approximately

$$\frac{Y - n\pi}{\sqrt{[n\pi(1 - \pi)]}} \sim N(0, 1)$$

This can be used to make inferences about the parameter π.

5.3 SAMPLING DISTRIBUTION FOR MAXIMUM LIKELIHOOD ESTIMATORS

Suppose that the log-likelihood function has a unique maximum at **b** and that this estimator **b** is near the true value of the parameter $\boldsymbol{\beta}$. The first-order Taylor approximation for the score vector $\mathbf{U}(\boldsymbol{\beta})$ about the point $\boldsymbol{\beta} = \mathbf{b}$ is given by

$$\mathbf{U}(\boldsymbol{\beta}) \cong \mathbf{U}(\mathbf{b}) + \boldsymbol{H}(\mathbf{b})(\boldsymbol{\beta} - \mathbf{b})$$

where $\boldsymbol{H}(\mathbf{b})$ denotes the matrix of second derivatives of the log-likelihood function evaluated at $\boldsymbol{\beta} = \mathbf{b}$. Asymptotically $\boldsymbol{H}$ is equal to its expected value which is related to the information matrix by

$$\mathcal{J} = E(\mathbf{U}\mathbf{U}^{\mathrm{T}}) = E(-\boldsymbol{H})$$

(see Appendix A). Therefore, for large samples

$$\mathbf{U}(\boldsymbol{\beta}) \cong \mathbf{U}(\mathbf{b}) - \mathcal{J}(\boldsymbol{\beta} - \mathbf{b})$$

But $\mathbf{U}(\mathbf{b}) = \mathbf{0}$ because $\mathbf{b}$ is the point at which the log-likelihood function is maximal and its derivatives are zero. Approximately, therefore,

$$(\mathbf{b} - \boldsymbol{\beta}) \cong \mathcal{J}^{-1}\mathbf{U}$$

provided that $\mathcal{J}$ is non-singular.

If $\mathcal{J}$ is regarded as constant then

$$E(\mathbf{b} - \boldsymbol{\beta}) \cong \mathcal{J}^{-1}E(\mathbf{U}) = \mathbf{0}$$

because $E(\mathbf{U}) = \mathbf{0}$ and so $\mathbf{b}$ is an unbiased estimator of $\boldsymbol{\beta}$ (at least asymptotically). The variance–covariance matrix for $\mathbf{b}$ is

$$E[(\mathbf{b} - \boldsymbol{\beta})(\mathbf{b} - \boldsymbol{\beta})^{\mathrm{T}}] \cong \mathcal{J}^{-1}E(\mathbf{U}\mathbf{U}^{\mathrm{T}})\mathcal{J}^{-1} = \mathcal{J}^{-1}$$

because $\mathcal{J} = E(\mathbf{U}\mathbf{U}^{\mathrm{T}})$ and $(\mathcal{J}^{-1})^{\mathrm{T}} = \mathcal{J}^{-1}$ since $\mathcal{J}$ is symmetric.

Thus for large samples

$$(\mathbf{b} - \boldsymbol{\beta})^{\mathrm{T}}\mathcal{J}(\mathbf{b} - \boldsymbol{\beta}) \sim \chi^2_p \qquad (5.2)$$

or, equivalently

$$\mathbf{b} - \boldsymbol{\beta} \sim N(\mathbf{0}, \mathcal{J}^{-1}) \qquad (5.3)$$

The statistic $(\mathbf{b} - \boldsymbol{\beta})^{\mathrm{T}}\mathcal{J}(\mathbf{b} - \boldsymbol{\beta})$ is sometimes called the **Wald statistic**. It is used to make inferences about $\boldsymbol{\beta}$.

For linear models with Normally distributed response variables results (5.2) and (5.3) are exact. This is shown in Example 5.3.

Example 5.3

Suppose that the response variables $Y_1, \ldots, Y_N$ are independently distributed with

$$Y_i \sim N(\mathbf{x}_i^{\mathrm{T}}\boldsymbol{\beta}, \sigma^2)$$

where σ^2 is a known constant. Let $\boldsymbol{X}$ be the $N \times p$ matrix consisting of the rows $\mathbf{x}_i^{\mathrm{T}}$ and suppose that $\boldsymbol{X}^{\mathrm{T}}\boldsymbol{X}$ is non-singular. In this case

$$E(Y_i) = \mu_i = \mathbf{x}_i^{\mathrm{T}}\boldsymbol{\beta}$$

The link function is the identity so, in the notation of Chapter 4, $\mu_i = \eta_i$ and therefore $\partial\mu_i/\partial\eta_i = 1$. Thus

1. From (4.8) the elements of $\mathscr{J}$ are given by

$$\mathscr{J}_{jk} = \frac{1}{\sigma^2} \sum_{i=1}^{N} x_{ij} x_{ik}$$

and so the information matrix $\mathscr{J}$ can be written as

$$\mathscr{J} = \frac{1}{\sigma^2} X^{\mathrm{T}} X \qquad (5.4)$$

2. From (4.9) $\boldsymbol{W}$ is the diagonal matrix with all elements equal to $1/\sigma^2$;
3. From (4.10) $\mathbf{z} = X\mathbf{b} + \mathbf{y} - X\mathbf{b} = \mathbf{y}$;
4. And so, from (4.11), the maximum likelihood estimator $\mathbf{b}$ is the solution of

$$X^{\mathrm{T}} X \mathbf{b} = X^{\mathrm{T}} \mathbf{y}$$

therefore

$$\mathbf{b} = (X^{\mathrm{T}} X)^{-1} X^{\mathrm{T}} \mathbf{y}$$

Since $\mathbf{b}$ is a linear combination of Normally distributed random variables $Y_1, \ldots, Y_N$ it, too, is Normally distributed. Also $\mathbf{b}$ is an unbiased estimator of $\boldsymbol{\beta}$ because

$$\begin{aligned} E(\mathbf{b}) &= (X^{\mathrm{T}} X)^{-1} X^{\mathrm{T}} E(\mathbf{y}) \\ &= (X^{\mathrm{T}} X)^{-1} X^{\mathrm{T}} X \boldsymbol{\beta} \qquad \text{because } E(\mathbf{y}) = X\boldsymbol{\beta} \\ &= \boldsymbol{\beta} \end{aligned}$$

To obtain the variance–covariance matrix for $\mathbf{b}$ we use

$$\begin{aligned} \mathbf{b} - \boldsymbol{\beta} &= (X^{\mathrm{T}} X)^{-1} X^{\mathrm{T}} \mathbf{y} - \boldsymbol{\beta} \\ &= (X^{\mathrm{T}} X)^{-1} X^{\mathrm{T}} (\mathbf{y} - X\boldsymbol{\beta}) \end{aligned}$$

therefore

$$\begin{aligned} E[(\mathbf{b} - \boldsymbol{\beta})(\mathbf{b} - \boldsymbol{\beta})^{\mathrm{T}}] &= (X^{\mathrm{T}} X)^{-1} X^{\mathrm{T}} E[(\mathbf{y} - X\boldsymbol{\beta})(\mathbf{y} - X\boldsymbol{\beta})^{\mathrm{T}}] X (X^{\mathrm{T}} X)^{-1} \\ &= \sigma^2 (X^{\mathrm{T}} X)^{-1} \end{aligned}$$

because $E[(\mathbf{y} - X\boldsymbol{\beta})(\mathbf{y} - X\boldsymbol{\beta})^{\mathrm{T}}]$ is the diagonal matrix with elements σ^2. But by (5.4) $X^{\mathrm{T}} X / \sigma^2 = \mathscr{J}$ so the variance–covariance matrix for $\mathbf{b}$ is $\mathscr{J}^{-1}$. Therefore the exact sampling distribution of $\mathbf{b}$ is $N(\boldsymbol{\beta}, \mathscr{J}^{-1})$, or equivalently,

$$(\mathbf{b} - \boldsymbol{\beta})^{\mathrm{T}} \mathscr{J} (\mathbf{b} - \boldsymbol{\beta}) \sim \chi_p^2$$

Thus results (5.2) and (5.3) are exact.

5.4 CONFIDENCE INTERVALS FOR THE MODEL PARAMETERS

Result (5.3), that the sampling distribution of the maximum likelihood estimator **b** is $N(\boldsymbol{\beta}, \mathcal{J}^{-1})$, can be used as follows:

1. To assess the reliability of the estimates b_j from the magnitudes of their standard errors

$$\text{s.e.}(b_j) = \sqrt{v_{jj}}$$

where v_{jj} is the jth term on the diagonal of the matrix $\mathcal{J}^{-1}$;
2. To calculate confidence intervals for individual parameters, for instance, an approximate 95% confidence interval for β_j is given by

$$b_j \pm 1.96\sqrt{v_{jj}}$$

3. To examine the correlations between the estimators using

$$\text{corr}\,(b_j, b_k) = \frac{v_{jk}}{\sqrt{v_{jj}}\sqrt{v_{kk}}}$$

Except for linear models with Normally distributed response variables the above results rely on large sample approximations. Also the information matrix $\mathcal{J}$ often depends on the parameters $\boldsymbol{\beta}$ so for practical applications we need to evaluate it at $\boldsymbol{\beta} = \mathbf{b}$, and occasionally $-\boldsymbol{H}(\mathbf{b})$ is used instead of $\mathcal{J}(\mathbf{b})$ as an estimate of $\mathcal{J}(\boldsymbol{\beta})$. For these reasons the results in this section are approximate, rather than exact, estimates of the quantities indicated.

Example 5.4 illustrates some uses of the sampling distribution of $\boldsymbol{\beta}$.

Example 5.4

In Section 4.5 we fitted to the data shown in Table 5.1 a model involving Poisson distributed responses Y_i with $E(Y_i) = \beta_1 + \beta_2 x_i$. The maximum likelihood estimates are $b_1 = 7.4516$ and $b_2 = 4.9353$. The inverse of the information matrix evaluated at **b** is

$$\mathcal{J}^{-1} = \begin{bmatrix} 0.7817 & 0.4166 \\ 0.4166 & 1.1863 \end{bmatrix}$$

Table 5.1 Poisson regression data

y_i	2	3	6	7	8	9	10	12	15
x_i	−1	−1	0	0	0	0	1	1	1

This shows that b_1 is somewhat more reliable than b_2 because its standard error is smaller – s.e. $(b_1) = 0.7817^{1/2} = 0.88$ compared to s.e. $(b_2) = 1.1863^{1/2} = 1.09$. The correlation coefficient for b_1 and b_2 is approximately

$$r = \frac{0.4166}{(0.7817)^{1/2}(1.1863)^{1/2}} \cong 0.43$$

An approximate 95% confidence interval for β_1 is given by $7.4516 \pm 1.96(0.7817)^{1/2}$, i.e. (5.72, 9.18).

5.5 ADEQUACY OF A MODEL

Suppose we are interested in assessing the adequacy of a model for describing a set of data. This can be done by comparing the likelihood under this model with the likelihood under the **maximal** or **full** model which is defined as follows:

1. The maximal model is a generalized linear model using the same distribution as the model of interest (e.g. both Normal or both binomial);
2. The maximal model has the same link function as the model of interest;
3. The number of parameters in the maximal model is equal to the total number of observations, N.

Because of 3. the maximal model can be regarded as providing a complete description of the data (at least for the assumed distribution).

The likelihood functions for the maximal model and the model of interest can be evaluated at the respective maximum likelihood estimates $\mathbf{b}_{\max}$ and $\mathbf{b}$ to obtain values $L(\mathbf{b}_{\max}; \mathbf{y})$ and $L(\mathbf{b}; \mathbf{y})$ respectively. If the model of interest describes the data well then $L(\mathbf{b}; \mathbf{y})$ should be approximately equal to $L(\mathbf{b}_{\max}; \mathbf{y})$. If the model is poor then $L(\mathbf{b}; \mathbf{y})$ will be much smaller than $L(\mathbf{b}_{\max}; \mathbf{y})$. This suggests the use of the generalized **likelihood ratio statistic**

$$\lambda = \frac{L(\mathbf{b}_{\max}; \mathbf{y})}{L(\mathbf{b}; \mathbf{y})}$$

as a measure of goodness of fit. Equivalently, the difference between the log-likelihood functions could be used

$$\log \lambda = l(\mathbf{b}_{\max}; \mathbf{y}) - l(\mathbf{b}; \mathbf{y})$$

Large values of $\log \lambda$ suggest that the model of interest is a poor description of the data. To determine the critical region for $\log \lambda$ we need to know its sampling distribution.

5.6 SAMPLING DISTRIBUTION FOR THE LOG-LIKELIHOOD STATISTIC

Suppose that the model of interest involves p parameters denoted by the parameter vector $\boldsymbol{\beta}$. A Taylor series approximation for $l(\boldsymbol{\beta}; \mathbf{y})$ can be obtained by expanding it about the maximum likelihood estimator $\mathbf{b}$

$$l(\boldsymbol{\beta}; \mathbf{y}) \cong l(\mathbf{b}; \mathbf{y}) + (\boldsymbol{\beta} - \mathbf{b})^{\mathrm{T}}\mathbf{U}(\mathbf{b}) + \tfrac{1}{2}(\boldsymbol{\beta} - \mathbf{b})^{\mathrm{T}}\boldsymbol{H}(\mathbf{b})(\boldsymbol{\beta} - \mathbf{b}) \quad (5.5)$$

where $\mathbf{U}(\mathbf{b})$ is the vector of scores $\partial l/\partial \beta_j$ evaluated at $\mathbf{b}$ and $\boldsymbol{H}(\mathbf{b})$ is the matrix of second derivatives

$$\frac{\partial^2 l}{\partial \beta_j \partial \beta_k}$$

evaluated at $\mathbf{b}$. From the definition of $\mathbf{b}$, $\mathbf{U}(\mathbf{b}) = \mathbf{0}$. Also for large samples $-\boldsymbol{H}(\mathbf{b})$ can be approximated by the information matrix because $\mathcal{J} = E[-\boldsymbol{H}]$. Thus (5.5) can be rearranged to obtain

$$l(\mathbf{b}; \mathbf{y}) - l(\boldsymbol{\beta}; \mathbf{y}) = \tfrac{1}{2}(\mathbf{b} - \boldsymbol{\beta})^{\mathrm{T}}\mathcal{J}(\mathbf{b} - \boldsymbol{\beta})$$

But from (5.2) $(\mathbf{b} - \boldsymbol{\beta})^{\mathrm{T}}\mathcal{J}(\mathbf{b} - \boldsymbol{\beta}) \sim \chi^2_p$, therefore

$$2[l(\mathbf{b}; \mathbf{y}) - l(\boldsymbol{\beta}; \mathbf{y})] \sim \chi^2_p \quad (5.6)$$

We use a test statistic based on this result to assess the fit of a model and to compare alternative models.

5.7 LOG-LIKELIHOOD RATIO STATISTIC (DEVIANCE)

We define the log-likelihood ratio statistic as

$$D = 2\log\lambda = 2[l(\mathbf{b}_{\max}; \mathbf{y}) - l(\mathbf{b}; \mathbf{y})] \quad (5.7)$$

Nelder and Wedderburn (1972) called this the (scaled) **deviance**. It can be rewritten as

$$\begin{aligned} D = 2\{&[l(\mathbf{b}_{\max}; \mathbf{y}) - l(\boldsymbol{\beta}_{\max}; \mathbf{y})] \\ &- [l(\mathbf{b}; \mathbf{y}) - l(\boldsymbol{\beta}; \mathbf{y})] \\ &+ [l(\boldsymbol{\beta}_{\max}; \mathbf{y}) - l(\boldsymbol{\beta}; \mathbf{y})]\} \end{aligned} \quad (5.8)$$

The first term in square brackets on the right-hand side of (5.8) has the χ^2_N distribution, by result (5.6) because the maximal model has N parameters. Similarly the second term has the χ^2_p distribution. The third term is a positive constant which will be near zero if the model with p parameters describes the data nearly as well as the maximal model does. Equation (5.8) indicates how the sampling distribution of D is derived. Roughly speaking, if the random variables defined by the first two terms are independent and the third (constant) term is near zero then

$$D \sim \chi^2_{N-p} \tag{5.9}$$

if the model is good. If the model is poor the third term on the right-hand side of (5.8) will be large and so D will be larger than predicted by χ^2_{N-p} (in fact D will approximately have a non-central chi-squared distribution).

In general, (5.9) does not provide a very good approximation for the sampling distribution. For Normal models, however, the result is exact. This is illustrated in the following example.

Example 5.5

Suppose that response variables $Y_1, \ldots, Y_N$ are independent and Normally distributed with means μ_i, which may differ, and a common variance σ^2. The log-likelihood function is

$$l(\boldsymbol{\beta}; \mathbf{y}) = -\frac{1}{2\sigma^2}\sum_{i=1}^{N}(y_i - \mu_i)^2 - \frac{1}{2}N\log(2\pi\sigma^2)$$

For the maximal model $E(Y_i) = \mu_i$ for each i so $\boldsymbol{\beta}$ has elements $\mu_1, \ldots, \mu_N$. By differentiating the log-likelihood function we obtain $\hat{\mu}_i = y_i$. Therefore

$$l(\mathbf{b}_{\max}; \mathbf{y}) = -\tfrac{1}{2}N\log(2\pi\sigma^2)$$

Now consider the model in which all the Y_i's have the same mean so that $\boldsymbol{\beta}$ has only one element μ. In this case $\hat{\mu} = \bar{y}$ and therefore

$$l(\mathbf{b}; \mathbf{y}) = -\frac{1}{2\sigma^2}\sum_{i=1}^{N}(y_i - \bar{y})^2 - \frac{1}{2}N\log(2\pi\sigma^2)$$

Hence from (5.7)

$$D = 2[l(\mathbf{b}_{\max}; \mathbf{y}) - l(\mathbf{b}; \mathbf{y})]$$

$$= \frac{1}{\sigma^2}\sum_{i=1}^{N}(y_i - \bar{y})^2$$

The statistic D is related to the sample variance

$$S^2 = \frac{1}{N-1}\sum_{i=1}^{N}(y_i - \bar{y})^2$$

by $D = (N-1)S^2/\sigma^2$. If the model with one common mean μ is correct then all the Y_i's have the distribution $N(\mu, \sigma^2)$ and so $(N-1)S^2/\sigma^2$ has the χ^2_{N-1} distribution. Thus $D \sim \chi^2_{N-1}$ and so result (5.9) is exact.

Example 5.6

Suppose that the response variables $Y_1, \ldots, Y_N$ are independent and have Poisson distributions with parameters λ_i. Their log-likelihood function is

$$l(\boldsymbol{\beta}; \mathbf{y}) = \sum y_i \log \lambda_i - \sum \lambda_i - \sum \log y_i!$$

For the maximal model the maximum likelihood estimates are $\hat{\lambda}_i = y_i$ so

$$l(\mathbf{b}_{\max}; \mathbf{y}) = \sum y_i \log y_i - \sum y_i - \sum \log y_i!$$

For the model in which all the Y_i's have the same parameter λ the maximum likelihood estimate is

$$\hat{\lambda} = \sum y_i / N = \bar{y}$$

so

$$l(\mathbf{b}; \mathbf{y}) = \sum y_i \log \bar{y} - N\bar{y} - \sum \log y_i!$$

Therefore

$$\begin{aligned} D &= 2[l(\mathbf{b}_{\max}; \mathbf{y}) - l(\mathbf{b}; \mathbf{y})] \\ &= 2\left[\sum y_i \log y_i - \sum y_i \log \bar{y}\right] \\ &= 2 \sum y_i \log (y_i/\bar{y}) \end{aligned}$$

From (5.9) if the one parameter model is correct $D \sim \chi^2_{N-1}$ approximately.

Example 5.7
Let $Y_1, \ldots, Y_N$ be independent Normally distributed response variables with means μ_i, which may differ, and a common variance σ^2. For the maximal model, as in Example 5.5,

$$l(\mathbf{b}_{\max}; \mathbf{y}) = -\tfrac{1}{2} N \log (2\pi\sigma^2)$$

Suppose that in the model of interest the means depend on p parameters $\boldsymbol{\beta}$ where $p < N$ (e.g. slope and intercept parameters if $\mu_i = \beta_1 + \beta_2 x_i$, or group means if $\mu_i = \beta_1$ for $i = 1, \ldots, m$ and $\mu_i = \beta_2$ for $i = m + 1, \ldots, N$). Let $\hat{\mu}_i$ denote the estimate of μ_i calculated from the maximum likelihood estimate $\mathbf{b}$. Then

$$l(\mathbf{b}; \mathbf{y}) = -\frac{1}{2\sigma^2} \sum_{i=1}^{N} (y_i - \hat{\mu}_i)^2 - \tfrac{1}{2} N \log (2\pi\sigma^2)$$

and hence

$$D = \frac{1}{\sigma^2} \sum_{i=1}^{N} (y_i - \hat{\mu}_i)^2$$

By (5.9), if the model is correct, $D \sim \chi^2_{N-p}$. The estimate $\hat{\mu}_i$ is often called the **fitted value** for Y_i. The value $(y_i - \hat{\mu}_i)$ is called the **residual**. Thus D is the sum of squares of the residuals divided by the nuisance parameter σ^2. The program GLIM gives the deviance $\sigma^2 D = \Sigma (y_i - \hat{\mu}_i)^2$.

5.8 ASSESSING GOODNESS OF FIT

The sampling distribution of the log-likelihood ratio statistic can be used to investigate the adequacy of a model by estimating D from the data and comparing the value with the appropriate chi-squared distribution. If the model is good we would expect the value of D to be near the middle of the distribution. This is easy to assess because the expected value of a random variable with the χ^2_m distribution is m. (This result can be readily obtained from the probability density function of the chi-squared distribution and is to be found in most elementary textbooks.) Thus if a model with p parameters provides a good description for a data set of N observations, so that $D \sim \chi^2_{N-p}$, we can expect

$$D \cong N - p \qquad (5.10)$$

For some distributions, such as the Poisson distribution (see Example 5.6), the value of D can be calculated directly from the fitted values and compared with its degrees of freedom to assess the goodness of fit.

For other distributions, such as the Normal distribution (see Example 5.7), D cannot be calculated directly because of the presence of nuisance parameters (e.g. σ^2). For the Normal distribution the program GLIM gives the value of the deviance $\sigma^2 D = \Sigma (y_i - \hat{\mu}_i)^2$ and also, corresponding to (5.10), gives a **scale parameter** which is an estimate of σ^2

$$\text{scale parameter} = \hat{\sigma}^2 = \text{deviance}/(N - p)$$

Generally (5.10) provides only a crude method for assessing the goodness of fit of a model. Other methods, especially those involving the examination of residuals, are often more useful; these are outlined in section 5.10.

Example 5.8
Consider again the Poisson regression model fitted to the data shown in Table 4.1. When the model

$$E(Y_i) = \beta_1 + \beta_2 x_i$$

with $p = 2$ parameters was fitted to the $N = 9$ observations, the GLIM output showed the scaled deviance (i.e. the log-likelihood ratio statistic) as

$$D = 1.8947$$

with 7 degrees of freedom (see section 4.7). This deviance is small relative to the degrees of freedom (in fact, it is below the lower 5% tail of the distribution) indicating that the model fits the data well – perhaps not surprising for such a small artificial set of data!

5.9 HYPOTHESIS TESTING

Hypotheses about the parameters $\boldsymbol{\beta}$ can be tested using the asymptotic sampling distribution for the estimator $\mathbf{b}$, namely $\mathbf{b} \sim N(\boldsymbol{\beta}, \mathcal{J}^{-1})$; or equivalently, the Wald statistic $(\mathbf{b} - \boldsymbol{\beta})^{\mathrm{T}}\mathcal{J}(\mathbf{b} - \boldsymbol{\beta})$ which has the χ^2_p distribution. Occasionally the score statistic $\mathbf{U}^{\mathrm{T}}\mathcal{J}^{-1}\mathbf{U}$, which also has the χ^2_p distribution, is used.

An alternative approach, which was illustrated in Chapter 2, consists of specifying each hypothesis in terms of a model and comparing the goodness-of-fit statistics for the competing models. The models to be compared need to have the same distribution and the same link function, i.e. only the number of parameters may differ. Consider the null hypothesis

$$H_0 : \boldsymbol{\beta} = \boldsymbol{\beta}_0 = \begin{bmatrix} \beta_1 \\ \vdots \\ \beta_q \end{bmatrix}$$

and a more general hypothesis

$$H_1 : \boldsymbol{\beta} = \boldsymbol{\beta}_1 = \begin{bmatrix} \beta_1 \\ \vdots \\ \beta_p \end{bmatrix} \quad \text{where } q < p < N$$

We can test H_0 against H_1 using the difference of log-likelihood ratio statistics

$$\begin{aligned} \Delta D = D_0 - D_1 &= 2[l(\mathbf{b}_{\max}; \mathbf{y}) - l(\mathbf{b}_0; \mathbf{y})] - 2[l(\mathbf{b}_{\max}; \mathbf{y}) - l(\mathbf{b}_1; \mathbf{y})] \\ &= 2[l(\mathbf{b}_1; \mathbf{y}) - l(\mathbf{b}_0; \mathbf{y})] \end{aligned}$$

If both models describe the data well then $D_0 \sim \chi^2_{N-q}$ and $D_1 \sim \chi^2_{N-p}$ so that $\Delta D \sim \chi^2_{p-q}$ (provided that certain independence conditions hold). If the value of ΔD is consistent with the χ^2_{p-q} distribution we would generally choose the model corresponding to H_0 because it is simpler.

If the value of ΔD is in the critical region (i.e. greater than the upper tail $100\alpha\%$ point of the χ^2_{p-q} distribution) then we would reject H_0 in favour of H_1 on the grounds that $\boldsymbol{\beta}_1$ provides a significantly better description of the data (even though it too may not fit the data particularly well). The sampling distribution of ΔD is usually much better approximated by the chi-squared distribution than is the sampling distribution of an individual log-likelihood ratio statistic D (i.e. result (5.9)).

For models involving the Normal distribution with a common variance σ^2 the log-likelihood ratio statistics depend on σ^2 so they cannot be calculated directly from the fitted values. This difficulty is overcome as follows. Let $\hat{\mu}_i(0)$ and $\hat{\mu}_i(1)$ denote the fitted values for the response variables Y_i under hypotheses H_0 and H_1 respectively. Then from Example 5.7

$$D_0 = \frac{1}{\sigma^2} \sum [y_i - \hat{\mu}_i(0)]^2$$

and

$$D_1 = \frac{1}{\sigma^2} \sum [y_i - \hat{\mu}_i(1)]^2$$

It is usual to assume that H_1 is correct so that $D_1 \sim \chi^2_{N-p}$. If H_0 is also correct then $D_0 \sim \chi^2_{N-q}$ and so $\Delta D = D_0 - D_1 \sim \chi^2_{p-q}$. If H_0 is not correct ΔD will have a non-central χ^2 distribution. To eliminate the term σ^2 we use the ratio

$$F = \frac{D_0 - D_1}{p - q} \bigg/ \frac{D_1}{N - p}$$

$$= \frac{\{\Sigma [y_i - \hat{\mu}_i(0)]^2 - \Sigma [y_i - \hat{\mu}_i(1)]^2\}/(p - q)}{\Sigma [y_i - \mu_i(1)]^2/(N - p)}$$

Thus F can be calculated directly from the fitted values. If H_0 is correct F will have the central $F_{p-q,N-p}$ distribution (at least approximately). If H_0 is not correct the value of F will be larger than expected from the $F_{p-q,N-p}$ distribution. These ideas are illustrated in Example 5.9.

Example 5.9

In Chapter 2 we used a model-fitting approach to test the null hypothesis that there was no difference in weight between plants grown under two different conditions. The data are shown in Table 5.2. Let Y_{jk} denote the weight of the kth plant grown under condition j, where $j = 1$ for control, $j = 2$ for treatment and $k = 1, \ldots, 10$. The weights

Table 5.2 Plant weights from two different growing conditions (from Table 2.1)

Control	4.17	5.58	5.18	6.11	4.50	4.61	5.17	4.53	5.33	5.14
Treatment	4.81	4.17	4.41	3.59	5.87	3.83	6.03	4.89	4.32	4.69

are assumed to be Normally distributed. The analysis consists of comparing

Model 1: $$Y_{jk} = \mu_j + e_{jk}$$

where μ_j is the mean weight for condition j, with

Model 0: $$Y_{jk} = \mu + e_{jk}$$

where the mean μ is the same for both conditions.

With GLIM these models can be fitted and compared to test the hypothesis that the means are the same.

```
? $units 20$
? $factor ct $
? $data y $
? $read
$REA? 4.17 5.58 5.18 6.11 4.50 4.61 5.17 4.53 5.33 5.14
$REA? 4.81 4.17 4.41 3.59 5.87 3.83 6.03 4.89 4.32 4.69
? $calc ct = %gl(2,10) $
? $yvar y $
? $error normal $
? $link identity $
? $fit $
deviance =   9.4175
      d.f. = 19
? $dis e $
             estimate              s.e.                  parameter
      1      4.846                 0.1574                1
      scale parameter taken as 0.4957
? $fit : + ct $
deviance =   9.4175
      d.f. = 19
deviance =   8.7292 (change = -0.6882)
      d.f. = 18       (change = -1      )
? $dis e $
             estimate              s.e.                  parameter
  1             5.403              0.4924                1
  2           -0.3710              0.3114                CT
  scale parameter taken as 0.4850
```

Model 0 is fitted first. It involves the single parameter $\beta = \mu$ and the (default) design matrix X is a column of 20 ones. For Model 1 the command % GL is used to define a second column for X with elements 1 for the control condition and 2 for the treatment condition so the group means are related to

$$\boldsymbol{\beta} = \begin{bmatrix} \beta_1 \\ \beta_2 \end{bmatrix} \text{ by } \mu_1 = \beta_1 + \beta_2 \text{ and } \mu_2 = \beta_1 + 2\beta_2$$

For Model 0 the estimated mean is $\hat{\mu} = 4.846$ and the log-likelihood ratio statistic is

$$D_0 = \sigma^2 \times \text{deviance} = 9.4175\sigma^2$$

with $20 - 1 = 19$ degrees of freedom. For Model 1 the means are

$$\hat{\mu}_1 = 5.403 - 0.371 = 5.032$$

and

$$\hat{\mu}_2 = 5.403 - 2 \times 0.371 = 4.661$$

and the log-likelihood ratio statistic is $D_1 = 8.7292\sigma^2$ with $20 - 2 = 18$ degrees of freedom. To test the hypothesis that the group means are the same we use

$$f = \frac{D_0 - D_1}{1} \Big/ \frac{D_1}{18} = 1.42$$

which is not significant compared to the $F_{1,18}$ distribution. So the data are consistent with the hypothesis that there is no difference in plant weight obtained under the two different growing conditions. Compare these results with section 2.2.

5.10 RESIDUALS

A goodness-of-fit statistic provides an overall measure of the adequacy of a model. The investigation of specific aspects of a model, however, is facilitated by the use of residuals.

Consider a Normal model in which the response variables Y_i are modelled by

$$Y_i = \mu_i + e_i$$

where the error terms e_i are assumed to be independent and all have the distribution $N(0, \sigma^2)$ and the expected values μ_i are a monotone function of linear combinations of elements of a parameter vector $\boldsymbol{\beta}$. For this model

$$(Y_i - \mu_i)/\sigma \sim N(0, 1)$$

The residual corresponding to Y_i is defined as $(y_i - \hat{\mu}_i)$ where $\hat{\mu}_i$ is the fitted value calculated from the maximum likelihood estimate **b**. The standardized residual is defined as

$$r_i = (y_i - \hat{\mu}_i)/\hat{\sigma}$$

where $\hat{\sigma}$ is an estimate of σ. Thus the standardized residuals will approximately have the distribution $N(0, 1)$; this result is not exact because there is some correlation among the r_i's (see section 6.8).

For other generalized linear models, residuals are defined, by analogy with the Normal case, as

$$r_i = (y_i - \hat{\mu}_i)/s_i$$

where s_i is the estimated standard deviation of the fitted value $\hat{\mu}_i$. These are the standardized residuals routinely printed out by GLIM. For the Poisson distribution, for example,

$$E(Y_i) = \text{var}(Y_i) = \lambda_i$$

so

$$r_i = \frac{y_i - \hat{\lambda}_i}{\sqrt{\hat{\lambda}_i}}$$

The Poisson residuals can be regarded as signed square roots of contributions to the Pearson goodness-of-fit statistic

$$\sum \frac{(o_i - e_i)^2}{e_i}$$

where o_i is the observed value y_i and e_i is the fitted value $\hat{\lambda}_i$ 'expected' from the model.

Numerous other definitions have been proposed for standardized residuals. Many of these are discussed by McCullagh and Nelder (1989), including the use of various transformations to improve the normality of the residuals. Others based on the signed square roots of contributions to the log-likelihood ratio statistic have been considered by Pregibon (1981) and Pierce and Schafer (1986).

Standardized residuals can be used to determine the adequacy of a model in the following ways:

1. They can be compared with the Normal distribution to assess the adequacy of the distributional assumptions in the model and to identify any unusual values. This can be done by inspecting their frequency distribution and looking for values beyond the likely range; for example, no more than 5% should be less than −1.96 or greater

than $+1.96$ and no more than 1% should be beyond ± 2.58;

2. A more sensitive method for assessing Normality is to use a **probability plot**. This involves plotting the ordered standardized residuals against the **normal scores** which are the expected values of the Normal order statistics. This can be done in MINITAB, for example, using the command NSCORE or in GLIM using suitable commands (see Exercise 5.4). The points should lie on a straight line and systematic deviations or outlying observations indicate departures from the model and should be investigated further;
3. The standardized residuals should be plotted against each of the covariates to see if the model adequately describes the effect of the covariate. If the model is adequate there should be no apparent pattern in the plot. If it is inadequate the points may display curvilinearity or some other systematic pattern which would suggest that additional or alternative terms should be included in the model to describe the effect of the covariate;
4. The residuals may be plotted against the fitted values and against other potential covariates. If there is any systematic pattern this suggests that additional covariates should be included in the model.

Excellent general discussions of the examination of residuals are given by Draper and Smith (1981), Belsley, Kuh and Welsch (1980) and Cook and Weisberg (1982). For further suggestions about residuals for generalized linear models see McCullagh and Nelder (1989) and Aitkin *et al.* (1989).

5.11 EXERCISES

5.1 Consider the single response variable Y with the binomial distribution $b(n, \pi)$.

(a) Find the Wald statistic $(\hat{\pi} - \pi)^{\mathrm{T}}\mathcal{J}(\hat{\pi} - \pi)$ where $\hat{\pi}$ is the maximum likelihood estimator of π and $\mathcal{J}$ is the information.

(b) Verify that the Wald statistic is the same as the score statistic $U^{\mathrm{T}}\mathcal{J}^{-1}U$ in this case (see Example 5.2).

(c) Find the log-likelihood statistic

$$2[l(\hat{\pi}; y) - l(\pi; y)]$$

(d) For large samples both the Wald/score statistic and the log-likelihood statistic approximately have the χ^2_1 distribution. For $n = 10$ and $y = 3$ use both statistics to assess the adequacy of the models:
(i) $\pi = 0.1$; (ii) $\pi = 0.3$; (iii) $\pi = 0.5$.
Do the two statistics lead to the same conclusions?

5.2 Find the log-likelihood ratio statistics for each of the following distributions. In each case consider a random sample $Y_1, \ldots, Y_N$ and compare the maximal model with the model indicated.

(a) Binomial distribution:

$$f(y_i; \pi_i) = \binom{n_i}{y_i} \pi_i^{y_i}(1 - \pi_i)^{n_i - y_i}$$

for the model with $\pi_i = \pi$ for all i.

(b) Exponential distribution: $f(y_i; \theta_i) = \theta_i \exp(-y_i\theta_i)$ for the model with $\theta_i = \theta$ for all i.

5.3 For the leukaemia survival data in Exercise 4.3:

(a) Obtain an approximate 95% confidence interval for the parameter β_1.

(b) By comparing the log-likelihood ratio statistics for two appropriate models, test the null hypothesis $\beta_2 = 0$ against the alternative hypothesis, $\beta_2 \neq 0$. What can you conclude about the use of initial white blood cell count as a predictor of survival time?

5.4 Calculate standardized residuals for model 0 in Example 5.9 and use a probability plot to investigate the assumption of Normality. (*Note:* In GLIM approximate normal scores can be obtained using the commands

```
$CALC L = %CU(1)
  : % N = L
  : NS = %ND((%CU(1) - 0.5)/%N) $
```

The standardized residuals should be sorted using the command SORT and then plotted against NS.)

5.5 For the exponential distribution, show that if $E(Y) = \theta$ then $\operatorname{var}(Y) = \theta^2$ and hence that standardized residuals can be defined by

$$r = (y - \hat{\theta})/\hat{\theta}$$

For the leukaemia survival data (Exercise 4.3) use such standardized residuals to investigate the adequacy of the model $E(Y_i) = \exp(\beta_1 + \beta_2 x_i)$.

6

Multiple regression

6.1 INTRODUCTION

We begin the discussion of particular generalized linear models by considering the simplest case, multiple linear regression which is usually represented by the equation

$$\mathbf{y} = \boldsymbol{X}\boldsymbol{\beta} + \mathbf{e} \tag{6.1}$$

where $\mathbf{y}$ is an $N \times 1$ response vector;

$\boldsymbol{X}$ is an $N \times p$ matrix of constants (mainly values of explanatory variables);

$\boldsymbol{\beta}$ is a $p \times 1$ vector of parameters;

$\mathbf{e}$ is an $N \times 1$ random vector whose elements are independent, and all have the Normal distribution $N(0, \sigma^2)$.

This is a generalized linear model with

$$E(Y_i) = \mu_i = \mathbf{x}_i^{\mathrm{T}}\boldsymbol{\beta}$$

where the elements Y_i of $\mathbf{y}$ are independent and Normally distributed; the link function $g(\mu)$ is the identity function because μ_i is already a linear combination of the parameters; and $\mathbf{x}_i^{\mathrm{T}}$ is the ith row of the matrix $\boldsymbol{X}$ in (6.1).

For multiple regression models the design matrix $\boldsymbol{X}$ must have linearly independent columns so that $\boldsymbol{X}^{\mathrm{T}}\boldsymbol{X}$ is non-singular (in Chapter 7 we consider models without this constraint).

First we consider several examples of regression. Then we review the theoretical results relating to regression models; most of these have been obtained already as examples and exercises in previous chapters. Finally we mention several practical aspects of the use of regression. More detailed discussion of multiple regression can be found, for example, in Draper and Smith (1981) or Kleinbaum, Kupper and Muller (1988).

6.2 EXAMPLES

Example 6.1 Simple linear regression

A straight-line relationship between a continuous response variable, which is assumed to be Normally distributed, and a single explanatory variable is modelled by

$$E(Y_i) = \beta_0 + \beta_1 x_i \qquad i = 1, \ldots, N$$

This corresponds to the model $E(\mathbf{y}) = \boldsymbol{X\beta}$ with

$$\mathbf{y} = \begin{bmatrix} Y_1 \\ \vdots \\ Y_N \end{bmatrix} \quad \boldsymbol{X} = \begin{bmatrix} 1 & x_1 \\ \vdots & \vdots \\ 1 & x_N \end{bmatrix} \quad \text{and} \quad \boldsymbol{\beta} = \begin{bmatrix} \beta_0 \\ \beta_1 \end{bmatrix}$$

The birthweight example in section 2.3 involved models of this kind.

Example 6.2 Multiple linear regression

The data in Table 6.1 show responses, percentages of total calories obtained from complex carbohydrates, for twenty male insulin-dependent diabetics who had been on a high-carbohydrate diet for six months. Compliance with the regime is thought to be related to age (in years), body weight (relative to 'ideal' weight for height) and other components of the diet, such as percentage of calories as protein. These other variables are treated as explanatory variables.

Table 6.1 Carbohydrate, age, weight and protein for twenty male insulin-dependent diabetics; for units, see text (data from K. Webb, personal communication)

Carbohydrate Y	Age x_1	Weight x_2	Protein x_3
33	33	100	14
40	47	92	15
37	49	135	18
27	35	144	12
30	46	140	15
43	52	101	15
34	62	95	14
48	23	101	17
30	32	98	15
38	42	105	14
50	31	108	17
51	61	85	19
30	63	130	19
36	40	127	20
41	50	109	15
42	64	107	16
46	56	117	18
24	61	100	13
35	48	118	18
37	28	102	14

If the response is linearly related to each of the covariates a suitable model is $E(\mathbf{y}) = X\boldsymbol{\beta}$ with

$$\mathbf{y} = \begin{bmatrix} Y_1 \\ \vdots \\ Y_N \end{bmatrix} \quad X = \begin{bmatrix} 1 & x_{11} & x_{12} & x_{13} \\ \vdots & \vdots & \vdots & \vdots \\ 1 & x_{N1} & x_{N2} & x_{N3} \end{bmatrix} \quad \text{and} \quad \boldsymbol{\beta} = \begin{bmatrix} \beta_0 \\ \vdots \\ \beta_3 \end{bmatrix}$$

where $N = 20$. We use these data for illustrative purposes later in this chapter.

Example 6.3 Polynomial regression

A curvilinear relationship between the response variable Y and a single explanatory variable x may be modelled by a polynomial

$$Y_i = \beta_0 + \beta_1 x_i + \beta_2 x_i^2 + \ldots + \beta_{p-1} x_i^{p-1} + e_i \tag{6.2}$$

This is a special case of model (6.1) with

$$\mathbf{y} = \begin{bmatrix} Y_1 \\ \vdots \\ Y_N \end{bmatrix} \quad X = \begin{bmatrix} 1 & x_1 & x_1^2 & \ldots & x_1^{p-1} \\ \vdots & \vdots & \vdots & & \vdots \\ 1 & x_N & x_N^2 & & x_N^{p-1} \end{bmatrix}$$

and

$$\boldsymbol{\beta} = \begin{bmatrix} \beta_0 \\ \vdots \\ \beta_{p-1} \end{bmatrix}$$

so that the powers of x_i are treated as distinct covariates. In practice it is inadvisable to use more than three or four terms in model (6.2) for several reasons:

1. The columns of X are closely related and if p is too large $X^{\mathrm{T}}X$ may be nearly singular (see section 6.9);
2. There is a danger of producing a model which fits the data very well within the range of observations but is poor for prediction outside this range;
3. Often it is implausible that the mechanism linking x and Y is really described by a high-order polynomial and an alternative formulation should be sought.

Example 6.4 Trigonometric regression

If the relationship between the response Y and an explanatory variable x is cyclic or **periodic** a suitable model might be

$$Y_i = \beta_0 + \beta_1 \cos \alpha_1 x_i + \beta_2 \sin \alpha_1 x_i + \beta_3 \cos \alpha_2 x_i + \beta_4 \sin \alpha_2 x_i \ldots + e_i \tag{6.3}$$

where the α_j's are known constants. In this case $E(\mathbf{y}) = \boldsymbol{X\beta}$ with

$$\boldsymbol{X} = \begin{bmatrix} 1 & \cos\alpha_1 x_1 & \sin\alpha_1 x_1 & \cos\alpha_2 x_1 & \sin\alpha_2 x_1 & \dots \\ \vdots & \vdots & \vdots & \vdots & \vdots & \\ 1 & \cos\alpha_1 x_N & \sin\alpha_1 x_N & \cos\alpha_2 x_N & \sin\alpha_2 x_N & \dots \end{bmatrix}$$

Trigonometric regression can be used to model seasonality in economic data, circadian rhythms and other periodic biological phenomena. For the same reasons as mentioned for polynomial regression, usually it is inadvisable to have too many terms on the right-hand side of (6.3).

6.3 MAXIMUM LIKELIHOOD ESTIMATION

If the response variables Y_i are independent and have the distributions $Y_i \sim N(\mathbf{x}_i^T\boldsymbol{\beta}, \sigma^2)$ then the log-likelihood function is

$$l = \frac{-1}{2\sigma^2}(\mathbf{y} - \boldsymbol{X\beta})^T(\mathbf{y} - \boldsymbol{X\beta}) - \frac{N}{2}\log(2\pi\sigma^2) \tag{6.4}$$

where

$$\mathbf{y} = \begin{bmatrix} Y_1 \\ \vdots \\ Y_N \end{bmatrix} \qquad \boldsymbol{X} = \begin{bmatrix} \mathbf{x}_1^T \\ \vdots \\ \mathbf{x}_N^T \end{bmatrix} \qquad \text{and} \quad \boldsymbol{\beta} = \begin{bmatrix} \beta_1 \\ \vdots \\ \beta_p \end{bmatrix}$$

From (6.4)

$$\mathbf{U} = \frac{\partial l}{\partial \boldsymbol{\beta}} = \frac{1}{\sigma^2}\boldsymbol{X}^T(\mathbf{y} - \boldsymbol{X\beta})$$

so the maximum likelihood estimator of $\boldsymbol{\beta}$ is given by the solution of $\boldsymbol{X}^T\boldsymbol{X}\mathbf{b} = \boldsymbol{X}^T\mathbf{y}$, i.e.

$$\mathbf{b} = (\boldsymbol{X}^T\boldsymbol{X})^{-1}\boldsymbol{X}^T\mathbf{y}$$

(since $\boldsymbol{X}^T\boldsymbol{X}$ is assumed to be non-singular – see section 6.1).

In Example 5.3 it is shown that $E(\mathbf{b}) = \boldsymbol{\beta}$ and $E[(\mathbf{b} - \boldsymbol{\beta})(\mathbf{b} - \boldsymbol{\beta})^T] = \sigma^2(\boldsymbol{X}^T\boldsymbol{X})^{-1}$. Also $\mathbf{b}$ is a linear combination of elements of the Normally distributed responses Y_i, so that

$$\mathbf{b} \sim N(\boldsymbol{\beta}, \sigma^2(\boldsymbol{X}^T\boldsymbol{X})^{-1}) \tag{6.5}$$

(More generally if $Y_i \sim N(\mathbf{x}_i^T\boldsymbol{\beta}, \sigma_i^2)$ then the maximum likelihood estimator of $\boldsymbol{\beta}$ is the solution of $\boldsymbol{X}^T\boldsymbol{V}^{-1}\boldsymbol{X}\mathbf{b} = \boldsymbol{X}^T\boldsymbol{V}^{-1}\mathbf{y}$ where $\boldsymbol{V}$ is the diagonal matrix with elements $v_{ii} = \sigma_i^2$ – see Exercise 4.5.)

For generalized linear models σ^2 is treated as a constant and it is not estimated so the distribution (6.5) is not fully determined. More conventionally $\boldsymbol{\beta}$ and σ^2 are estimated simultaneously to give the maximum likelihood estimates

$$\mathbf{b} = (X^{\mathrm{T}}X)^{-1}X^{\mathrm{T}}\mathbf{y} \quad \text{and} \quad \tilde{\sigma}^2 = \frac{1}{N}(\mathbf{y} - X\mathbf{b})^{\mathrm{T}}(\mathbf{y} - X\mathbf{b})$$

However, it can be shown that the estimator $\tilde{\sigma}^2$ is not unbiased, in fact $E(\tilde{\sigma}^2) = (N - p)\sigma^2/N$, so that an unbiased estimator of σ^2 can be defined by

$$\hat{\sigma}^2 = \frac{1}{N - p}(\mathbf{y} - X\mathbf{b})^{\mathrm{T}}(\mathbf{y} - X\mathbf{b}) \tag{6.6}$$

Using results (6.5) and (6.6) confidence intervals and hypothesis tests for $\boldsymbol{\beta}$ can be derived.

6.4 LEAST SQUARES ESTIMATION

If $E(\mathbf{y}) = X\boldsymbol{\beta}$ and $E[(\mathbf{y} - X\boldsymbol{\beta})(\mathbf{y} - X\boldsymbol{\beta})^{\mathrm{T}}] = V$ where V is known we can obtain the least squares estimator of $\boldsymbol{\beta}$ without making any further assumptions about the distribution of $\mathbf{y}$. We minimize

$$S_w = (\mathbf{y} - X\boldsymbol{\beta})^{\mathrm{T}}V^{-1}(\mathbf{y} - X\boldsymbol{\beta})$$

The solution of

$$\frac{\partial S_w}{\partial \boldsymbol{\beta}} = -2X^{\mathrm{T}}V^{-1}(\mathbf{y} - X\boldsymbol{\beta}) = \mathbf{0}$$

is

$$\mathbf{b} = (X^{\mathrm{T}}V^{-1}X)^{-1}X^{\mathrm{T}}V^{-1}\mathbf{y}$$

(provided that the matrix inverses exist) – see section 4.3. In particular if the elements of $\mathbf{y}$ are independent and have a common variance then

$$\mathbf{b} = (X^{\mathrm{T}}X)^{-1}X^{\mathrm{T}}\mathbf{y}$$

Thus for regression models with Normal errors, maximum likelihood and least squares estimators are the same.

6.5 LOG-LIKELIHOOD RATIO STATISTIC

For a maximal model in which

$$\boldsymbol{\beta}_{\text{max}} = \begin{bmatrix} \beta_1 \\ \vdots \\ \beta_N \end{bmatrix}$$

without loss of generality we can take X as the unit matrix so that $\mathbf{b}_{\text{max}} = \mathbf{y}$ and by substitution in (6.4)

$$l(\mathbf{b}_{\text{max}}; \mathbf{y}) = -\tfrac{1}{2}N\log(2\pi\sigma^2)$$

For any other model $E(\mathbf{y}) = X\boldsymbol{\beta}$ involving p parameters

$$\boldsymbol{\beta} = \begin{bmatrix} \beta_1 \\ \vdots \\ \beta_p \end{bmatrix}$$

where $p < N$, let $\mathbf{b}$ denote the maximum likelihood estimator. Then the log-likelihood ratio statistic is

$$\begin{aligned} D &= 2[l(\mathbf{b}_{\max}; \mathbf{y}) - l(\mathbf{b}; \mathbf{y})] \\ &= \frac{1}{\sigma^2}(\mathbf{y} - X\mathbf{b})^{\mathrm{T}}(\mathbf{y} - X\mathbf{b}) \\ &= \frac{1}{\sigma^2}(\mathbf{y}^{\mathrm{T}}\mathbf{y} - 2\mathbf{b}^{\mathrm{T}}X^{\mathrm{T}}\mathbf{y} + \mathbf{b}^{\mathrm{T}}X^{\mathrm{T}}X\mathbf{b}) \\ &= \frac{1}{\sigma^2}(\mathbf{y}^{\mathrm{T}}\mathbf{y} - \mathbf{b}^{\mathrm{T}}X^{\mathrm{T}}\mathbf{y}) \end{aligned}$$

because $X^{\mathrm{T}}X\mathbf{b} = X^{\mathrm{T}}\mathbf{y}$.

If the model is correct then $D \sim \chi^2_{N-p}$, otherwise D has the non-central chi-squared distribution with $N - p$ degrees of freedom (from section 5.7).

The statistic D is not completely determined when σ^2 is unknown. As illustrated in previous chapters, for hypothesis testing we overcome this difficulty by using appropriately defined ratios of log-likelihood ratio statistics.

As in section 5.9, consider a null hypothesis H_0 and an alternative hypothesis H_1 which can be specified in terms of models with parameters

$$H_0 : \boldsymbol{\beta} = \boldsymbol{\beta}_0 = \begin{bmatrix} \beta_1 \\ \vdots \\ \beta_q \end{bmatrix} \quad \text{and} \quad H_1 : \boldsymbol{\beta} = \boldsymbol{\beta}_1 = \begin{bmatrix} \beta_1 \\ \vdots \\ \beta_p \end{bmatrix}$$

where $q < p < N$. Let X_0 and X_1 denote the corresponding design matrices, $\mathbf{b}_0$ and $\mathbf{b}_1$ the maximum likelihood estimators and D_0 and D_1 the log-likelihood ratio statistics. We test H_0 against H_1 using

$$\begin{aligned} \Delta D = D_0 - D_1 &= \frac{1}{\sigma^2}[(\mathbf{y}^{\mathrm{T}}\mathbf{y} - \mathbf{b}_0^{\mathrm{T}}X_0^{\mathrm{T}}\mathbf{y}) - (\mathbf{y}^{\mathrm{T}}\mathbf{y} - \mathbf{b}_1^{\mathrm{T}}X_1^{\mathrm{T}}\mathbf{y})] \\ &= \frac{1}{\sigma^2}(\mathbf{b}_1^{\mathrm{T}}X_1^{\mathrm{T}}\mathbf{y} - \mathbf{b}_0^{\mathrm{T}}X_0^{\mathrm{T}}\mathbf{y}) \end{aligned}$$

We assume that H_1 is correct so that $D_1 \sim \chi^2_{N-p}$. If H_0 is also correct then $D_0 \sim \chi^2_{N-q}$, otherwise D_0 has a non-central chi-squared distribution with $N - q$ degrees of freedom. Thus if H_0 is correct $D_0 - D_1 \sim \chi^2_{p-q}$ and so

$$f = \frac{D_0 - D_1}{p - q} \bigg/ \frac{D_1}{N - p} = \frac{\mathbf{b}_1^{\mathrm{T}} X_1^{\mathrm{T}} \mathbf{y} - \mathbf{b}_0^{\mathrm{T}} X_0^{\mathrm{T}} \mathbf{y}}{p - q} \bigg/ \frac{\mathbf{y}^{\mathrm{T}} \mathbf{y} - \mathbf{b}_1^{\mathrm{T}} X_1^{\mathrm{T}} \mathbf{y}}{N - p} \sim F_{p-q, N-p}$$

If H_0 is not correct, f has a non-central F distribution. Therefore, values of f which are large relative to the $F_{p-q,N-p}$ distribution provide evidence against H_0. This test for H_0 is often set out as shown in Table 6.2.

Table 6.2 Analysis of variance table

Source of variation	Degrees of freedom	Sum of squares	Mean square
Model with $\boldsymbol{\beta}_0$	q	$\mathbf{b}_0^{\mathrm{T}} X_0^{\mathrm{T}} \mathbf{y}$	
Improvement due to model with $\boldsymbol{\beta}_1$	$p - q$	$\mathbf{b}_1^{\mathrm{T}} X_1^{\mathrm{T}} \mathbf{y}_1 - \mathbf{b}_0^{\mathrm{T}} X_0^{\mathrm{T}} \mathbf{y}$	$\dfrac{\mathbf{b}_1^{\mathrm{T}} X_1^{\mathrm{T}} \mathbf{y} - \mathbf{b}_0^{\mathrm{T}} X_0^{\mathrm{T}} \mathbf{y}}{p - q}$
Residual	$N - p$	$\mathbf{y}^{\mathrm{T}} \mathbf{y} - \mathbf{b}_1^{\mathrm{T}} X_1^{\mathrm{T}} \mathbf{y}$	$\dfrac{\mathbf{y}^{\mathrm{T}} \mathbf{y} - \mathbf{b}_1^{\mathrm{T}} X_1^{\mathrm{T}} \mathbf{y}}{N - p}$
Total	N	$\mathbf{y}^{\mathrm{T}} \mathbf{y}$	

This form of analysis is one of the major tools for hypothesis testing using regression models (and also for analysis of variance – see Chapter 7). It depends on the assumption that the most general model fitted, in this case $E(\mathbf{y}) = X_1 \boldsymbol{\beta}_1$, describes the data well so that the corresponding statistic D has the central chi-squared distribution. This assumption should be checked, for example, by examining the residuals (see section 6.8.).

Another less rigorous comparison of goodness of fit between two models is provided by R^2, the square of the multiple correlation coefficient.

6.6 MULTIPLE CORRELATION COEFFICIENT AND R^2

If $\mathbf{y} = X\boldsymbol{\beta} + \mathbf{e}$ and the elements of $\mathbf{e}$ are independent with $E(e_i) = 0$ and $\operatorname{var}(e_i) = \sigma^2$ for all i then the least squares criterion is

$$S = \sum_{i=1}^{N} e_i^2 = \mathbf{e}^{\mathrm{T}} \mathbf{e} = (\mathbf{y} - X\boldsymbol{\beta})^{\mathrm{T}} (\mathbf{y} - X\boldsymbol{\beta})$$

The minimum value of S for the model is

$$\hat{S} = (\mathbf{y} - X\mathbf{b})^{\mathrm{T}} (\mathbf{y} - X\mathbf{b}) = \mathbf{y}^{\mathrm{T}} \mathbf{y} - \mathbf{b}^{\mathrm{T}} X^{\mathrm{T}} \mathbf{y}$$

(from section 6.4). This can be used as a measure of the fit of the model.

The value of S is compared with the fit of the simplest or **minimal model** $E(Y_i) = \mu$ for all i. This model can be written in the general form $E(\mathbf{y}) = X\boldsymbol{\beta}$ if $\boldsymbol{\beta} = [\mu]$ and $X = \mathbf{1}$, where $\mathbf{1}$ is the $N \times 1$ vector of ones. Therefore $X^{\mathrm{T}}X = N$, $X^{\mathrm{T}}\mathbf{y} = \Sigma y_i$ and $\mathbf{b} = \hat{\mu} = \bar{y}$. The corresponding value of the least squares criterion is

$$\hat{S}_0 = \mathbf{y}^{\mathrm{T}}\mathbf{y} - N\bar{y}^2 = \sum_{i=1}^{N}(y_i - \bar{y})^2$$

Thus $\hat{S}_0$ is proportional to the variance of the observations and it is regarded as the 'worst' possible value of S.

Any other model, with value $\hat{S}$, is assessed relative to the minimal model using $\hat{S}_0$. The difference

$$\hat{S}_0 - \hat{S} = \mathbf{b}^{\mathrm{T}}X^{\mathrm{T}}\mathbf{y} - N\bar{y}^2$$

is the improvement in fit due to the model $E(\mathbf{y}) = X\boldsymbol{\beta}$.

$$R^2 = \frac{\hat{S}_0 - \hat{S}}{\hat{S}_0} = \frac{\mathbf{b}^{\mathrm{T}}X^{\mathrm{T}}\mathbf{y} - N\bar{y}^2}{\mathbf{y}^{\mathrm{T}}\mathbf{y} - N\bar{y}^2}$$

is interpreted as the proportion of the total variation in the data which is explained by the model.

If the model does not describe the data any better than the minimal model then $\hat{S}_0 \cong \hat{S}$ so $R^2 \cong 0$. If the maximal model with N parameters is used then X is the $N \times N$ unit matrix $\boldsymbol{I}$, so that $\mathbf{b} = \mathbf{y}$ and $\mathbf{b}^{\mathrm{T}}X^{\mathrm{T}}\mathbf{y} = \mathbf{y}^{\mathrm{T}}\mathbf{y}$ and hence $R^2 = 1$, corresponding to a 'perfect' fit. In general $0 < R^2 < 1$. The square root of R^2 is called the **multiple correlation coefficient**.

A disadvantage of using R^2 as a measure of goodness of fit is that its sampling distribution is not readily determined. Also its value is not adjusted for the number of parameters used in the fitted model.

The use of R^2 and hypothesis testing based on the log-likelihood ratio statistic are illustrated in the following numerical example.

6.7 NUMERICAL EXAMPLE

We use the carbohydrate data in Table 6.1 and begin by fitting the model

$$E(Y_i) = \beta_0 + \beta_1 x_{i1} + \beta_2 x_{i2} + \beta_3 x_{i3} \tag{6.7}$$

in which carbohydrate Y is linearly related to age x_1, weight x_2 and protein x_3. If

$$\mathbf{y} = \begin{bmatrix} Y_1 \\ \vdots \\ Y_N \end{bmatrix} \quad X = \begin{bmatrix} 1 & x_{11} & x_{12} & x_{13} \\ \vdots & \vdots & \vdots & \vdots \\ 1 & x_{N1} & x_{N2} & x_{N3} \end{bmatrix} \quad \text{and} \quad \boldsymbol{\beta} = \begin{bmatrix} \beta_0 \\ \vdots \\ \beta_3 \end{bmatrix}$$

then for these data

$$X^{\mathrm{T}}\mathbf{y} = \begin{bmatrix} 752 \\ 34\,596 \\ 82\,270 \\ 12\,105 \end{bmatrix}$$

and

$$X^{\mathrm{T}}X = \begin{bmatrix} 20 & 923 & 2\,214 & 318 \\ 923 & 45\,697 & 102\,003 & 14\,780 \\ 2214 & 102\,003 & 250\,346 & 35\,306 \\ 318 & 14\,780 & 35\,306 & 5\,150 \end{bmatrix}$$

Therefore the solution of $X^{\mathrm{T}}X\mathbf{b} = X^{\mathrm{T}}\mathbf{y}$ is

$$\mathbf{b} = \begin{bmatrix} 36.9601 \\ -0.1137 \\ -0.2280 \\ 1.9577 \end{bmatrix}$$

and

$$(X^{\mathrm{T}}X)^{-1} = \begin{bmatrix} 4.8158 & -0.0113 & -0.0188 & -0.1362 \\ -0.0113 & 0.0003 & 0.0000 & -0.0004 \\ -0.0188 & 0.0000 & 0.0002 & -0.0002 \\ -0.1362 & -0.0004 & -0.0002 & 0.0114 \end{bmatrix}$$

correct to four decimal places. Also $\mathbf{y}^{\mathrm{T}}\mathbf{y} = 29\,368$, $N\bar{y}^2 = 28\,275.2$ and $\mathbf{b}^{\mathrm{T}}X^{\mathrm{T}}\mathbf{y} = 28800.337$ so that $R^2 = 0.48$, i.e. 48% of the total variation in the data is explained by model (6.7). Using (6.6) to obtain an unbiased estimator of σ^2 we get $\hat{\sigma}^2 = 35.479$ and hence the standard errors for elements of **b** which are shown in Table 6.3.

To illustrate the use of the log-likelihood ratio statistic we test the

Table 6.3 Estimates for model (6.7)

Term	Estimate b_j	Standard error[a]
Constant	36.960	13.071
Coefficient for age	−0.114	0.109
Coefficient for weight	−0.228	0.083
Coefficient for protein	1.958	0.635

[a]Values calculated using more significant figures for $(X^{\mathrm{T}}X)^{-1}$ than shown above.

hypothesis, H_0, that the response does not depend on age, i.e. $\beta_1 = 0$. The corresponding model is

$$E(Y_i) = \beta_0 + \beta_2 x_{i2} + \beta_3 x_{i3} \tag{6.8}$$

The matrix X for this model is obtained from the previous one by omitting the second column so that

$$\mathbf{X}^T\mathbf{y} = \begin{bmatrix} 752 \\ 82\,270 \\ 12\,105 \end{bmatrix} \qquad \mathbf{X}^T\mathbf{X} = \begin{bmatrix} 20 & 2214 & 318 \\ 2214 & 250\,346 & 35\,306 \\ 318 & 35\,306 & 5150 \end{bmatrix}$$

and hence

$$\mathbf{b} = \begin{bmatrix} 33.130 \\ -0.222 \\ 1.824 \end{bmatrix}$$

For model (6.8) $\mathbf{b}^T\mathbf{X}^T\mathbf{y} = 28\,761.978$ so that $R^2 = 0.445$, i.e. 44.5% of the variation is explained by the model. The significance test for H_0 is summarized in Table 6.4. The value $f = 38.36/35.48 = 1.08$ is not significant compared with the $F_{1,16}$ distribution so the data provide no evidence against H_0, i.e. the response appears to be unrelated to age.

Table 6.4 Analysis of variance table comparing models (6.7) and (6.8)

Source variation	Degrees of freedom	Sum of squares	Mean square
Model (6.8)	3	28 761.978	
Improvement due to model (6.7)	1	38.359	38.36
Residual	16	567.663	35.48
Total	20	29 368.000	

6.8 RESIDUAL PLOTS

For the regression model (6.1) we assume that the error terms e_i are independent and Normally distributed, all with mean 0 and variance σ^2, and that they do not vary systematically with $\mathbf{y}$ or elements of $\mathbf{X}$. These assumptions can be checked by examining the residuals

$$\hat{e}_i = y_i - \mathbf{x}_i^T\mathbf{b}$$

If the model is correct then $\hat{\mathbf{e}}$, the vector of residuals, has the properties $E(\hat{\mathbf{e}}) = \mathbf{0}$ and

$$E(\hat{\mathbf{e}}\hat{\mathbf{e}}^{\mathrm{T}}) = E(\mathbf{y}\mathbf{y}^{\mathrm{T}}) - XE(\mathbf{b}\mathbf{b}^{\mathrm{T}})X^{\mathrm{T}} = \sigma^2[I - X(X^{\mathrm{T}}X)^{-1}X^{\mathrm{T}}]$$

where I is the unit matrix. So the **standardized residuals** are defined by

$$r_i = \frac{\hat{e}_i}{\hat{\sigma}(1 - p_{ii})^{1/2}}$$

where p_{ii} is the ith element on the diagonal of the **projection or hat matrix** $\boldsymbol{P} = X(X^{\mathrm{T}}X)^{-1}X^{\mathrm{T}}$.

Probability plots of these residuals can be used to test the assumption of Normality. Also the standardized residuals should be very nearly uncorrelated so substantial serial correlations between them may indicate misspecification of the model; therefore it is often worth while to check for serial correlation (e.g. using the Durbin–Watson test).

The standardized residuals should also be plotted against the fitted values $\hat{y}_i = \mathbf{x}_i^{\mathrm{T}}\mathbf{b}$ and against each of the explanatory variables. Patterns in these plots can indicate misspecifications of the model and they can be used to identify any unusual observations which may have a strong influence on the value of **b** and on the goodness of fit of the model.

An excellent discussion of the examination of residuals for multiple regression models is given in Chapter 3 of the book by Draper and Smith (1981) while Aitkin *et al.* (1989) describe model checking with GLIM using residuals.

6.9 ORTHOGONALITY

In the numerical example in section 6.7 the parameters β_0, β_2 and β_3 occurred in both models (6.7) and (6.8) but their estimates differed when the models were fitted to the data. Also the test of the hypothesis that $\beta_1 = 0$ depends on which terms were included in the model. For example, the analysis of variance table comparing the models

$$E(Y_i) = \beta_0 + \beta_1 x_{i1} + \beta_3 x_{i3} \quad (6.9)$$

and

$$E(Y_i) = \beta_0 + \beta_3 x_{i3} \quad (6.10)$$

(which do not include $\beta_2 x_{i2}$) will differ from Table 6.4 comparing models (6.7) and (6.8) – see Exercise 6.3(c).

Usually estimates, confidence intervals and hypothesis tests depend on which covariates are included in the model. An exception is when the matrix X is **orthogonal**, i.e. it can be partitioned into components $X_1, X_2, \ldots, X_m$ corresponding to submodels of interest

$$X = [X_1, \ldots, X_m] \qquad \text{for } m \leqslant p$$

with the property that $X_j^T X_k = O$, a matrix of zeros, for each $j \neq k$. Let

$$\boldsymbol{\beta} = \begin{bmatrix} \boldsymbol{\beta}_1 \\ \vdots \\ \boldsymbol{\beta}_m \end{bmatrix}$$

be the corresponding partition of the parameters, then

$$E(\mathbf{y}) = X\boldsymbol{\beta} = X_1\boldsymbol{\beta}_1 + \ldots + X_m\boldsymbol{\beta}_m$$

and $X^T X$ is the block diagonal matrix

$$X^T X = \begin{bmatrix} X_1^T X_1 & & O \\ & \ddots & \\ O & & X_m^T X_m \end{bmatrix} \quad \text{also} \quad X^T \mathbf{y} = \begin{bmatrix} X_1^T \mathbf{y} \\ \vdots \\ X_m^T \mathbf{y} \end{bmatrix}$$

where O is used to indicate that the remaining elements of the matrix are zeros. Therefore the estimates $\mathbf{b}_j = (X_j^T X_j)^{-1} X_j^T \mathbf{y}$ are unaltered by the omission or inclusion of other components in the model and also

$$\mathbf{b}^T X^T \mathbf{y} = \mathbf{b}_1^T X_1^T \mathbf{y} + \ldots + \mathbf{b}_m^T X_m^T \mathbf{y}$$

Also the hypotheses

$$H_1 : \boldsymbol{\beta}_1 = \mathbf{0}, \ldots, \qquad H_m : \boldsymbol{\beta}_m = \mathbf{0}$$

can be tested independently as shown in Table 6.5.

Table 6.5 Multiple hypothesis tests when the design matrix X is orthogonal

Source of variation	Degrees of freedom	Sum of squares
Model corresponding to H_1	p_1	$\mathbf{b}_1^T X_1^T \mathbf{y}$
$\vdots$	$\vdots$	$\vdots$
Model corresponding to H_m	p_m	$\mathbf{b}_m^T X_m^T \mathbf{y}$
Residual	$N - \sum_{j=1}^{m} p_j$	$\mathbf{y}^T\mathbf{y} - \mathbf{b}^T X^T \mathbf{y}$
Total	N	$\mathbf{y}^T\mathbf{y}$

Unfortunately the benefits of orthogonality can only be exploited if X can be designed to have this property. This may be possible if the elements of X are dummy variables representing factor levels (see Chapter 7) or if polynomial regression is performed using **orthogonal polynomials** (these are specially constructed polynomials such that the columns of X corresponding to successively higher powers of the explanatory variable are orthogonal – see Draper and Smith, 1981, section 5.6).

6.10 COLLINEARITY

If the explanatory variables are closely related to one another the columns of X may nearly be linearly dependent so that $X^T X$ is nearly singular. In this case the equation $X^T X\mathbf{b} = X^T\mathbf{y}$ is said to be **ill-conditioned** and the solution **b** will be unstable in the sense that small changes in the data may cause large changes in **b**. Also at least some of the elements of $\sigma^2(X^T X)^{-1}$ will be large corresponding to large variance and covariance estimates of **b**. Thus careful inspection of the matrix $(X^T X)^{-1}$ may reveal the presence of collinearity.

The resolution of the problem is more difficult. It may require extra information from the substantive area from which the data came, an alternative specification of the model or some other non-computational approach. In addition various computational techniques, such as **ridge regression**, have been proposed for handling this problem. Detailed discussions of collinearity are given, for example, in the books by Belsley, Kuh and Welsch (1980) and Draper and Smith (1981).

A particular difficulty with collinearity occurs in the selection of some subset of the explanatory variables which 'best' describes the data. If two variables are highly correlated it may be impossible, on statistical grounds alone, to determine which should be included in a model.

6.11 MODEL SELECTION

Many applications of regression involve large numbers of explanatory variables and an important issue is to identify a subset of these variables which provides a good and parsimonious model for the response. The usual procedure is sequentially to add or delete terms from the model; this is called **stepwise regression**. Unless the variables are orthogonal this involves considerable computation and multiple testing of related hypotheses (with associated difficulties in interpreting significance levels). For further consideration of these problems the reader is referred to any of the standard textbooks on regression (e.g. Draper and Smith, 1981).

6.12 NON-LINEAR REGRESSION

The term **non-linear regression** is used for two types of models. The first is models of the form $E(Y) = g(\mathbf{x}^T\boldsymbol{\beta})$ which are generalized linear models provided that the distribution of Y is a member of the exponential family and the link function g is monotone. An example is Holliday's (1960) equation for plant yield

$$E(Y) = \frac{1}{\beta_0 + \beta_1 x + \beta_2 x^2}$$

where Y is the yield per plant and x is a measure of plant density. If Y is assumed to be Normally distributed then the methods of Chapters 4 and 5 can be used for estimation and inference.

The second type of non-linear regression model is of the form, $E(Y) = g(\mathbf{x}, \boldsymbol{\beta})$ where g is intrinsically non-linear in the parameters, for example, the logistic growth model

$$E(Y) = \frac{\beta_0}{1 + \beta_1 \exp(\beta_2 x)}$$

For these cases iterative estimation methods can be used analogous to those in Chapter 4 (see Charnes, Frome and Yu, 1976; Ratkowsky and Dolby, 1975). However the distributional results which hold for generalized linear models do not apply, for example the sampling distributions of the estimators may be seriously non-Normal (Ratkowsky, 1983).

6.13 EXERCISES

6.1 Table 6.6 shows the average apparent per capita consumption of sugar (in kg per year) in Australia, as refined sugar and in manufactured foods (from Australian Bureau of Statistics, publication 4306.0).

Table 6.6

Sugar consumption	1936–39	1946–49	1956–59	1966–69	1976–79	1983–86
As refined sugar	32.0	31.2	27.0	21.0	14.9	9.9
In manufactured foods	16.3	23.1	23.6	27.7	34.6	34.5

(a) Plot sugar consumption against time separately for refined sugar and sugar in manufactured food. (*Note:* The first five time-periods are ten years apart but the last is only seven years from the previous one.) Fit simple linear regression models to summarize these data. Calculate 95% confidence intervals for the average annual change in consumption for each form of sugar.

(b) Calculate the total average sugar consumption for each period and plot these data against time. Using suitable models test the hypothesis that total sugar consumption did not change over time.

6.2 Table 6.7 shows response of a grass/legume pasture system to various quantities of phosphorus fertilizer (data from D. F. Sinclair; results reported in Sinclair and Probert, 1986). The total yield, of grass and legume together, and amount of phosphorus are both given in kilograms per hectare. Find a suitable model for describing the relationship between yield and quantity of fertilizer. To do this:

(a) Plot yield against phosphorus to obtain an approximately linear relationship – you may need to try several transformations of either or both variables in order to achieve approximate linearity.
(b) Use the results of (a) to specify a possible model. Fit the model.
(c) Calculate the standardized residuals for the model and use appropriate plots to check for any systematic effects which might suggest alternative models and to investigate the validity of any assumptions made.

Table 6.7

Phosphorus	Yield	Phosphorus	Yield	Phosphorus	Yield
0	1753.9	15	3107.7	10	2400.0
40	4923.1	30	4415.4	5	2861.6
50	5246.2	50	4938.4	40	3723.0
5	3184.6	5	3046.2	30	4892.3
10	3538.5	0	2553.8	40	4784.6
30	4000.0	10	3323.1	20	3184.6
15	4184.6	40	4461.5	0	2723.1
40	4692.3	20	4215.4	50	4784.6
20	3600.0	40	4153.9	15	3169.3

6.3 Analyse the carbohydrate data in Table 6.1 using an appropriate computer program (or, preferably, repeat the analyses using several different regression programs and compare the results).

(a) Plot the responses y against each of the explanatory variables x_1, x_2 and x_3 to see if y appears to be linearly related to them.
(b) Fit the full model (6.7) and examine the residuals to assess the adequacy of the model and the assumptions.
(c) Fit models (6.9) and (6.10) and use these to test the hypothesis: $\beta_1 = 0$. Compare your results with Table 6.4.

Table 6.8

Cholesterol	Age	Body mass	Cholesterol	Age	Body mass
5.94	52	20.7	6.48	65	26.3
4.71	46	21.3	8.83	76	22.7
5.86	51	25.4	5.10	47	21.5
6.52	44	22.7	5.81	43	20.7
6.80	70	23.9	4.65	30	18.9
5.23	33	24.3	6.82	58	23.9
4.97	21	22.2	6.28	78	24.3
8.78	63	26.2	5.15	49	23.8
5.13	56	23.3	2.92	36	19.6
6.74	54	29.2	9.27	67	24.3
5.95	44	22.7	5.57	42	22.0
5.83	71	21.9	4.92	29	22.5
5.74	39	22.4	6.72	33	24.1
4.92	58	20.2	5.57	42	22.7
6.69	58	24.4	6.25	66	27.3

6.4 It is well known that the concentration of cholesterol in blood serum increases with age but it is less clear whether cholesterol level is also associated with body weight. Table 6.8 shows for thirty women serum cholesterol (millimoles per litre), age (years) and body mass index (weight divided by height squared, where weight was measured in kilograms and height in metres). Use multiple regression to test whether serum cholesterol is associated with body mass index when age is already included in the model.

7

Analysis of variance and covariance

7.1 INTRODUCTION

This chapter concerns linear models of the form

$$\mathbf{y} = \boldsymbol{X\beta} + \mathbf{e} \qquad \text{with } \mathbf{e} \sim N(\mathbf{0}, \sigma^2\boldsymbol{I})$$

where $\mathbf{y}$ and $\mathbf{e}$ are random vectors of length N, $\boldsymbol{X}$ is an $N \times p$ matrix of constants, $\boldsymbol{\beta}$ is a vector of p parameters and $\boldsymbol{I}$ is the unit matrix. These models differ from the regression models of Chapter 6 in that the **design matrix** $\boldsymbol{X}$ consists entirely of dummy variables for analysis of variance (ANOVA) or dummy variables and measured covariates for analysis of covariance (ANCOVA). Since the choice of dummy variables is to some extent arbitrary, a major consideration is the optimal choice of $\boldsymbol{X}$. The main questions addressed by analysis of variance and covariance involve comparisons of means. Traditionally the emphasis is on hypothesis testing rather than estimation or prediction.

In this book we only consider **fixed effects models** in which the levels of factors are regarded as fixed so that $\boldsymbol{\beta}$ is a vector of constants. We do not consider **random effects models** where the factor levels are regarded as a random selection from a population of possible levels and $\boldsymbol{\beta}$ is treated as a vector of random variables. The problem of estimating variances for the elements of $\boldsymbol{\beta}$ in random effects models, also called **variance components models**, is discussed by McCullagh and Nelder (1989) in the framework of generalized linear models. Also the elements of the response vector $\mathbf{y}$ are assumed to be independent and therefore we do not consider situations involving **repeated** or **longitudinal** measurements on the same experimental units because then the observations are likely to be correlated.

Wider coverage of analysis of variance and covariance is provided by any of the conventional books on the subject, for example Hocking (1985) or Winer (1971).

7.2 BASIC RESULTS

Since the random components **e** in ANOVA and ANCOVA models are assumed to be Normally distributed many of the results obtained in Chapter 6 apply here too. For instance the log-likelihood function is

$$l = -\frac{1}{2\sigma^2}(\mathbf{y} - \boldsymbol{X\beta})^{\mathrm{T}}(\mathbf{y} - \boldsymbol{X\beta}) - \frac{N}{2}\log(2\pi\sigma^2)$$

so the maximum likelihood (or least squares) estimator **b** is the solution of the normal equations

$$\boldsymbol{X}^{\mathrm{T}}\boldsymbol{X}\mathbf{b} = \boldsymbol{X}^{\mathrm{T}}\mathbf{y} \tag{7.1}$$

In ANOVA models there are often more parameters than there are independent equations in $E(\mathbf{y}) = \boldsymbol{X\beta}$ therefore $\boldsymbol{X}^{\mathrm{T}}\boldsymbol{X}$ is singular and there is no unique solution of (7.1). In this case $\boldsymbol{\beta}$ is said to be not **estimable** or not **identifiable**. To obtain a particular solution extra equations are used so that **b** is the solution of

$$\boldsymbol{X}^{\mathrm{T}}\boldsymbol{X}\mathbf{b} = \boldsymbol{X}^{\mathrm{T}}\mathbf{y}$$

and (7.2)

$$\boldsymbol{C}\mathbf{b} = \mathbf{0}$$

In anticipation of the need for the extra equations $\boldsymbol{C}\mathbf{b} = \mathbf{0}$, the model $E(\mathbf{y}) = \boldsymbol{X\beta}$ often includes the **constraint equations** $\boldsymbol{C\beta} = \mathbf{0}$. The minimum value of the term $(\mathbf{y} - \boldsymbol{X\beta})^{\mathrm{T}}(\mathbf{y} - \boldsymbol{X\beta})$, however, is unique and it is obtained using any solution of (7.1), so the value of $(\mathbf{y} - \boldsymbol{X}\mathbf{b})^{\mathrm{T}}(\mathbf{y} - \boldsymbol{X}\mathbf{b})$ does not depend on the choice of the constraint equations (see Exercise 7.5). Other properties of **b** do depend on the choice of $\boldsymbol{C}$ as illustrated in the numerical examples in sections 7.3 and 7.4.

As shown in section 6.5, for a maximal model with N parameters the maximum likelihood estimator is $\mathbf{b}_{\max} = \mathbf{y}$ and so

$$l(\mathbf{b}_{\max}; \mathbf{y}) = -\frac{N}{2}\log(2\pi\sigma^2)$$

For any other model with p parameters and maximum likelihood estimator **b**, the log-likelihood ratio statistic is

$$\begin{aligned} D = 2[l(\mathbf{b}_{\max}; \mathbf{y}) - l(\mathbf{b}; \mathbf{y})] &= \frac{1}{\sigma^2}(\mathbf{y} - \boldsymbol{X}\mathbf{b})^{\mathrm{T}}(\mathbf{y} - \boldsymbol{X}\mathbf{b}) \\ &= \frac{1}{\sigma^2}(\mathbf{y}^{\mathrm{T}}\mathbf{y} - \mathbf{b}^{\mathrm{T}}\boldsymbol{X}^{\mathrm{T}}\mathbf{y}) \end{aligned} \tag{7.3}$$

If the model is correct $D \sim \chi^2_{N-p}$, otherwise D has a non-central chi-squared distribution. As with regression models D is not completely determined when σ^2 is unknown so that hypotheses are tested by comparing appropriate ratios of log-likelihood ratio statistics and using the F-distribution.

7.3 ONE-FACTOR ANOVA

The data in Table 7.1 are an extension of the plant weight example of Chapter 2. An experiment is conducted to compare yields (as measured by dried weight of plants) obtained under a control and two different treatment conditions. Thus the response, plant weight, depends on one factor, growing condition, with three levels – control, treatment A and treatment B. We are interested in whether response means differ among the three groups.

Table 7.1 Plant weights from three different growing conditions

Control	4.17	5.58	5.18	6.11	4.50	4.61	5.17	4.53	5.33	5.14
Treatment A	4.81	4.17	4.41	3.59	5.87	3.83	6.03	4.89	4.32	4.69
Treatment B	6.31	5.12	5.54	5.50	5.37	5.29	4.92	6.15	5.80	5.26

More generally, if experimental units are randomly allocated to groups corresponding to J levels of a factor, this is called a **completely randomized experimental design** and the data can be set out as in Table 7.2. The responses can be written as the column vector

$$\mathbf{y} = [Y_{11}, \ldots, Y_{1n_1}, Y_{21}, \ldots, Y_{2n_2}, \ldots, Y_{Jn_J}]^{\mathrm{T}}$$

of length $N = \Sigma_{j=1}^{J} n_j$. $Y_{11}, Y_{12}, \ldots, Y_{1n_1}$ have the same expected value so they are replicates. More generally all responses Y_{jk} with the same j are replicates. For simplicity in this discussion we only consider the case when all the samples are of the same size, i.e. $n_j = K$ so that $N = JK$.

Table 7.2 Data for one-factor ANOVA with J levels of the factor and unequal sample sizes

Factor level	Responses		Totals
A_1	Y_{11}	$Y_{12} \ldots Y_{1n_1}$	$Y_{1\cdot}$
A_1	Y_{21}	$Y_{22} \ldots Y_{2n_2}$	$Y_{2\cdot}$
$\vdots$	$\vdots$	$\vdots$	$\vdots$
A_J	Y_{J1}	$Y_{J2} \ldots Y_{Jn_J}$	$Y_{J\cdot}$

We consider three different formulations of the model corresponding to the hypothesis that the response means differ for different levels of the factor. The simplest version of the model is

$$E(Y_{jk}) = \mu_j \qquad j = 1, \ldots, J \tag{7.4}$$

In terms of elements Y_i of the vector $\mathbf{y}$, this can be written as

$$E(Y_i) = \sum_{j=1}^{J} x_{ij}\mu_j \qquad i = 1, \ldots, N$$

where $x_{ij} = 1$ if response Y_i corresponds to level A_j and $x_{ij} = 0$ otherwise. Thus

$$E(\mathbf{y}) = X\boldsymbol{\beta} \qquad \text{with } \boldsymbol{\beta} = \begin{bmatrix} \mu_1 \\ \mu_2 \\ \vdots \\ \mu_J \end{bmatrix} \quad \text{and} \quad X = \begin{bmatrix} \mathbf{1} & \mathbf{0} & \cdots & \mathbf{0} \\ \mathbf{0} & \mathbf{1} & & \\ \cdot & \cdot & \boldsymbol{O} & \\ \cdot & \boldsymbol{O} & \cdot & \mathbf{0} \\ \mathbf{0} & & & \mathbf{1} \end{bmatrix}$$

where $\mathbf{0}$ and $\mathbf{1}$ are vectors of length K of zeros and ones respectively, and $\boldsymbol{O}$ indicates that the remaining terms of the matrix are all zeros. Then $X^{\mathrm{T}}X$ is the $J \times J$ diagonal matrix

$$X^{\mathrm{T}}X = \begin{bmatrix} K & & & \\ & \ddots & \boldsymbol{O} & \\ & & K & \\ & \boldsymbol{O} & & \ddots & \\ & & & & K \end{bmatrix} \quad \text{and} \quad X^{\mathrm{T}}\mathbf{y} = \begin{bmatrix} Y_{1.} \\ Y_{2.} \\ \vdots \\ Y_{J.} \end{bmatrix}$$

so that

$$\mathbf{b} = \frac{1}{K}\begin{bmatrix} Y_{1.} \\ Y_{2.} \\ \vdots \\ Y_{J.} \end{bmatrix} = \begin{bmatrix} \bar{y}_1 \\ \bar{y}_2 \\ \vdots \\ \bar{y}_J \end{bmatrix}$$

and

$$\mathbf{b}^{\mathrm{T}}X^{\mathrm{T}}\mathbf{y} = \frac{1}{K}\sum_{j=1}^{J} Y_{j.}^2$$

The fitted values are $\hat{\mathbf{y}} = [\bar{y}_1, \bar{y}_1, \ldots, \bar{y}_1, \bar{y}_2, \ldots, \bar{y}_J]^{\mathrm{T}}$. The disadvantage of this simple formulation of the model is that it cannot be extended to more than one factor. For generalizability, we need to specify the model so that parameters for levels and combinations of levels of factors reflect differential effects beyond some average response.

The second model is one such formulation

$$E(Y_{jk}) = \mu + \alpha_j \qquad j = 1, \ldots, J$$

where μ is the average effect for all levels and α_j is an additional effect due to level A_j. For this parametrization there are $J + 1$ parameters.

$$\boldsymbol{\beta} = \begin{bmatrix} \mu \\ \alpha_1 \\ \vdots \\ \alpha_J \end{bmatrix} \qquad X = \begin{bmatrix} \mathbf{1} & \mathbf{1} & \mathbf{0} & \dots & \mathbf{0} \\ \mathbf{1} & \mathbf{0} & \mathbf{1} & & \\ \vdots & & & O & \\ \vdots & O & & & \cdot \\ \mathbf{1} & & & & \mathbf{1} \end{bmatrix}$$

where **0** and **1** are vectors of length K. Thus

$$X^{\mathrm{T}}\mathbf{y} = \begin{bmatrix} Y_{..} \\ Y_{1.} \\ \vdots \\ Y_{J.} \end{bmatrix} \quad \text{and} \quad X^{\mathrm{T}}X = \begin{bmatrix} N & K & \dots & K \\ K & K & & \\ \vdots & & \cdot & O \\ \vdots & O & & \\ K & & & K \end{bmatrix}$$

The first row of the $(J+1) \times (J+1)$ matrix $X^{\mathrm{T}}X$ is the sum of the remaining rows so $X^{\mathrm{T}}X$ is singular and there is no unique solution of the normal equations $X^{\mathrm{T}}X\mathbf{b} = X^{\mathrm{T}}\mathbf{y}$. The general solution can be written as

$$\mathbf{b} = \begin{bmatrix} \hat{\mu} \\ \hat{\alpha}_1 \\ \vdots \\ \hat{\alpha}_J \end{bmatrix} = \frac{1}{K}\begin{bmatrix} 0 \\ Y_{1.} \\ \vdots \\ Y_{J.} \end{bmatrix} - \lambda \begin{bmatrix} -1 \\ 1 \\ \vdots \\ 1 \end{bmatrix}$$

where λ is an arbitrary constant. It is traditional to impose the additional **sum-to-zero** constraint

$$\sum_{j=1}^{J} \alpha_j = 0$$

so that

$$\frac{1}{K}\sum_{j=1}^{J} Y_{j.} - J\lambda = 0$$

and hence

$$\lambda = \frac{1}{JK}\sum_{j=1}^{J} Y_{j.} = \frac{Y_{..}}{N}$$

This gives the solution

$$\hat{\mu} = \frac{Y_{..}}{N} \quad \text{and} \quad \hat{\alpha}_j = \frac{Y_{j.}}{K} - \frac{Y_{..}}{N} \qquad \text{for } j = 1, \dots, J$$

Hence

$$\mathbf{b}^{\mathrm{T}}X^{\mathrm{T}}\mathbf{y} = \frac{Y_{..}^2}{N} + \sum_{j=1}^{J} Y_{j.}\left(\frac{Y_{j.}}{K} - \frac{Y_{..}}{N}\right) = \frac{1}{K}\sum_{j=1}^{J} Y_{j.}^2$$

which is the same as for the first version of the model and the fitted values $\hat{\mathbf{y}} = [\bar{y}_1, \bar{y}_1, \dots, \bar{y}_J]^{\mathrm{T}}$ are also the same. Sum-to-zero constraints are used in most standard statistical computing programs.

A third version of the model is $E(Y_{jk}) = \mu + \alpha_j$ with the constraint that $\alpha_1 = 0$. Thus μ represents the effect of the first level and α_j measures the difference between the first and jth levels of the factor. This is called a **corner-point parametrization**; it is used in the program GLIM. For this version there are J parameters.

$$\boldsymbol{\beta} = \begin{bmatrix} \mu \\ \alpha_2 \\ \vdots \\ \alpha_J \end{bmatrix} \quad \text{also} \quad X = \begin{bmatrix} 1 & 0 & \dots & 0 \\ 1 & 1 & & \\ \vdots & \cdot & & O \\ \vdots & & \cdot & \\ \vdots & O & & \cdot \\ 1 & & & 1 \end{bmatrix}$$

so

$$X^T\mathbf{y} = \begin{bmatrix} Y_{..} \\ Y_{2.} \\ \vdots \\ Y_{J.} \end{bmatrix} \quad \text{and} \quad X^TX = \begin{bmatrix} N & K & \dots & K \\ K & K & & O \\ \vdots & & \ddots & \\ K & O & & K \end{bmatrix}$$

The $J \times J$ matrix X^TX is non-singular so there is a unique solution

$$\mathbf{b} = \frac{1}{K}\begin{bmatrix} Y_{1.} \\ Y_{2.} - Y_{1.} \\ \vdots \\ Y_{J.} - Y_{1.} \end{bmatrix}$$

Also

$$\mathbf{b}^TX^T\mathbf{y} = \frac{1}{K}\left[Y_{..}Y_{1.} + \sum_{j=2}^{J} Y_{j.}(Y_{j.} - Y_{1.})\right] = \frac{1}{K}\sum_{j=1}^{J} Y_{j.}^2$$

and the fitted values $\hat{\mathbf{y}} = [\bar{y}_1, \bar{y}_1, \dots, \bar{y}_J]^T$ are the same as before.

Thus although the three specifications of the model differ, the value of $\mathbf{b}^TX^T\mathbf{y}$ and hence

$$D_1 = \frac{1}{\sigma^2}(\mathbf{y}^T\mathbf{y} - \mathbf{b}^TX^T\mathbf{y}) = \frac{1}{\sigma^2}\left[\sum_{j=1}^{J}\sum_{k=1}^{K} Y_{jk}^2 - \frac{1}{K}\sum_{j=1}^{J} Y_{j.}^2\right]$$

is the same in each case.

These three versions of the model all correspond to the hypothesis H_1 that the response means for each level may differ. To compare this with the null hypothesis H_0 that the means are all equal, we consider the model $E(Y_{jk}) = \mu$ so that $\boldsymbol{\beta} = [\mu]$ and X is a vector of N ones. Then $X^TX = N$, $X^T\mathbf{y} = Y_{..}$ and hence $\mathbf{b} = \hat{\mu} = Y_{..}/N$ so that $\mathbf{b}^TX^T\mathbf{y} = Y_{..}^2/N$ and

$$D_0 = \frac{1}{\sigma^2}\left[\sum_{j=1}^{J}\sum_{k=1}^{K} Y_{jk}^2 - \frac{Y_{..}^2}{N}\right]$$

To test H_0 against H_1 we assume that H_1 is correct so that $D_1 \sim \chi^2_{N-J}$. If, in addition, H_0 is correct then $D_0 \sim \chi^2_{N-1}$, otherwise D_0 has a non-central chi-squared distribution. Thus if H_0 is correct

$$D_0 - D_1 = \frac{1}{\sigma^2}\left[\frac{1}{K}\sum_{j=1}^{J} Y_{j.}^2 - \frac{1}{N} Y_{..}^2\right] \sim \chi^2_{J-1}$$

and so

$$f = \frac{D_0 - D_1}{J-1} \bigg/ \frac{D_1}{N-J} \sim F_{J-1,N-J}$$

If H_0 is not correct then f is likely to be larger than predicted from the $F_{J-1,N-J}$ distribution. Conventionally this hypothesis test is set out in an ANOVA table.

For the plant weight data

$$\frac{Y_{..}^2}{N} = 772.0599, \qquad \frac{1}{K}\sum_{j=1}^{J} Y_{j.}^2 = 775.8262$$

so

$$D_0 - D_1 = 3.7663/\sigma^2$$

and

$$\sum_{j=1}^{J}\sum_{k=1}^{K} Y_{jk}^2 = 786.3183 \quad \text{so} \quad D_1 = 10.4921/\sigma^2$$

Hence the hypothesis test is summarized in Table 7.3.

Table 7.3 ANOVA table for plant weight data in Table 7.1

Source of variation	Degrees of freedom	Sum of squares	Mean square	f
Mean	1	772.0599		
Between treatments	2	3.7663	1.883	4.85
Residual	27	10.4921	0.389	
Total	30	786.3183		

Since $f = 4.85$ is significant at the 5% level when compared with the $F_{2,27}$ distribution, we conclude that the group means differ.

To investigate this result further it is convenient to use the first version of the model, $E(Y_{jk}) = \mu_j$. The estimated means are

$$\mathbf{b} = \begin{bmatrix} \hat{\mu}_1 \\ \hat{\mu}_2 \\ \hat{\mu}_3 \end{bmatrix} = \begin{bmatrix} 5.032 \\ 4.661 \\ 5.526 \end{bmatrix}$$

If we use the estimate

$$\hat{\sigma}^2 = \frac{1}{N - J}(\mathbf{y} - X\mathbf{b})^{\mathrm{T}}(\mathbf{y} - X\mathbf{b}) = \frac{1}{N - J}(\mathbf{y}^{\mathrm{T}}\mathbf{y} - \mathbf{b}^{\mathrm{T}}X^{\mathrm{T}}\mathbf{y})$$

(equation 6.6), we obtain $\hat{\sigma}^2 = 10.4921/27 = 0.389$ (i.e. the residual mean square in Table 7.3). The variance–covariance matrix of **b** is $\hat{\sigma}^2(X^{\mathrm{T}}X)^{-1}$ where

$$X^{\mathrm{T}}X = \begin{bmatrix} 10 & 0 & 0 \\ 0 & 10 & 0 \\ 0 & 0 & 10 \end{bmatrix}$$

so the standard error of each element of **b** is $\sqrt{0.389/10} = 0.197$. Now it can be seen that the significant effect is due to the mean for treatment B, $\hat{\mu}_3 = 5.526$, being significantly larger than the other two means.

7.4 TWO-FACTOR ANOVA WITH REPLICATION

Consider the fictitious data in Table 7.4 in which factor A (with $J = 3$ levels) and factor B (with $K = 2$ levels) are **crossed** so that there are JK subclasses formed by all combinations of A and B levels. In each subclass there are $L = 2$ observations or **replicates**.

Table 7.4 Fictitious data for two-factor ANOVA with equal numbers of observations in each subclass

Levels of factor A	Levels of factor B		
	B_1	B_2	Total
A_1	6.8, 6.6	5.3, 6.1	24.8
A_2	7.5, 7.4	7.2, 6.5	28.6
A_3	7.8, 9.1	8.8, 9.1	34.8
Total	45.2	43.0	88.2

The main hypotheses are:

H_{I}: there are no interaction effects, i.e. the effects of A and B are additive;

H_{A}: there are no differences in response associated with different levels of factor A;

H_{B}: there are no differences in response associated with different levels of factor B.

Thus we need to consider a **saturated model** and three **reduced models** formed by omitting various terms from the full model.

1. The saturated model is

$$E(Y_{jkl}) = \mu + \alpha_j + \beta_k + (\alpha\beta)_{jk} \tag{7.5}$$

where the terms $(\alpha\beta)_{jk}$ correspond to **interaction effects** and α_j and β_k to **main effects** of the factors;
2. The **additive model** is

$$E(Y_{jkl}) = \mu + \alpha_j + \beta_k \tag{7.6}$$

This is compared to the saturated model to test hypothesis H_{I}.
3. The model formed by omitting effects due to B is

$$E(Y_{jkl}) = \mu + \alpha_j \tag{7.7}$$

This is compared to the additive model to test hypothesis H_{B}.
4. The model formed by omitting effects due to A is

$$E(Y_{jkl}) = \mu + \beta_j \tag{7.8}$$

This is compared to the additive model to test hypothesis H_{A}.

The models (7.5)–(7.8) have too many parameters; for instance replicates in the same subclass have the same expected value so there can be at most JK independent expected values but the saturated model has $1 + J + K + JK = (J + 1)(K + 1)$ parameters. To overcome this difficulty (which leads to the singularity of $\boldsymbol{X}^{\mathrm{T}}\boldsymbol{X}$) we can impose the extra constraints

$$\alpha_1 + \alpha_2 + \alpha_3 = 0 \qquad \beta_1 + \beta_2 = 0$$

$$(\alpha\beta)_{11} + (\alpha\beta)_{12} = 0 \qquad (\alpha\beta)_{21} + (\alpha\beta)_{22} = 0 \qquad (\alpha\beta)_{31} + (\alpha\beta)_{32} = 0$$

$$(\alpha\beta)_{11} + (\alpha\beta)_{21} + (\alpha\beta)_{31} = 0$$

(the remaining condition $(\alpha\beta)_{12} + (\alpha\beta)_{22} + (\alpha\beta)_{32} = 0$ follows from the last four equations). These are the conventional constraint equations for ANOVA. Alternatively we can take

$$\alpha_1 = \beta_1 = (\alpha\beta)_{11} = (\alpha\beta)_{12} = (\alpha\beta)_{21} = (\alpha\beta)_{31} = 0$$

as the corner-point constraints. In either case the numbers of (linearly) independent parameters are: 1 for μ, $J - 1$ for the α_j's, $K - 1$ for the β_k's, and $(J - 1)(K - 1)$ for the $(\alpha\beta)_{jk}$'s.

Details of fitting all four models using either the sum-to-zero constraints or the corner-point constraints are given in Appendix C.

For models (7.5)–(7.8) the estimates $\mathbf{b}$ depend on the choice of constraints and dummy variables. However, the fitted values $\hat{\mathbf{y}} = \boldsymbol{X}\mathbf{b}$

are the same for all specifications of the models and so are the values of $\mathbf{b}^T X^T \mathbf{y}$ and $\sigma^2 D = \mathbf{y}^T\mathbf{y} - \mathbf{b}^T X^T \mathbf{y}$. For these data $\mathbf{y}^T\mathbf{y} = 664.1$ and the other results are summarized in Table 7.5 (the subscripts S, I, A, B and M refer to the saturated model, the models corresponding to H_I, H_A and H_B, and the model with only the overall mean respectively).

Table 7.5 Summary of calculations for data in Table 7.4

Terms in model	Number of parameters	Degrees of freedom	$\mathbf{b}^T X^T \mathbf{y}$	Deviance $\sigma^2 D = \mathbf{y}^T\mathbf{y} - \mathbf{b}^T X^T \mathbf{y}$
$\mu + \alpha_j + \beta_k + (\alpha\beta)_{jk}$	6	6	662.6200	$\sigma^2 D_S = 1.4800$
$\mu + \alpha_j + \beta_k$	4	8	661.4133	$\sigma^2 D_I = 2.6867$
$\mu + \alpha_j$	3	9	661.0100	$\sigma^2 D_B = 3.0900$
$\mu + \beta_k$	2	10	648.6733	$\sigma^2 D_A = 15.4267$
μ	1	11	648.2700	$\sigma^2 D_M = 15.8300$

To test H_I we assume that the saturated model is correct so that $D_S \sim \chi^2_6$ (there are 6 degrees of freedom because there are $N = 12$ observations and the model has $JK = 6$ independent parameters). If H_I is also correct then $D_I \sim \chi^2_8$ so that $D_I - D_S \sim \chi^2_2$ and

$$f = \frac{D_I - D_S}{2} \bigg/ \frac{D_S}{6} \sim F_{2,6}$$

The value of

$$f = \frac{2.6867 - 1.48}{2\sigma^2} \bigg/ \frac{1.48}{6\sigma^2} = 2.45$$

is not significant so the data do not provide evidence against H_I. Since H_I is not rejected we proceed to test H_A and H_B. For H_B we consider the difference in fit between the models

$$E(Y_{jkl}) = \mu + \alpha_j \quad \text{and} \quad E(Y_{jkl}) = \mu + \alpha_j + \beta_k$$

i.e. $D_B - D_I$ and compare this with D_S using

$$f = \frac{D_B - D_I}{1} \bigg/ \frac{D_S}{6} = \frac{3.09 - 2.6867}{\sigma^2} \bigg/ \frac{1.48}{6\sigma^2} = 1.63$$

which is not significant compared to the $F_{1,6}$ distribution, suggesting that there are no differences due to levels of factor B. The corresponding test for H_A gives $f = 25.82$ which is significant compared with the $F_{2,6}$ distribution. Thus we conclude that the response means are affected only by differences in the levels of factor A.

For these analyses we have assumed that the saturated model provides a good description of the data and so D_S has a central chi-squared distribution. Therefore D_S was used in the denominator for all the F-tests. This corresponds to the conventional ANOVA approach (Table 7.6). For these data, however, it could be argued that as we do not reject H_I the additive model describes the data about as well as the saturated model and it is simpler so that $D_I \sim \chi_8^2$ should be used in the F-tests for H_A and H_B. Use of D_I is arguably more consistent with the model-fitting approach to data analysis.

Table 7.6 ANOVA table for data in Table 7.4

Source of variation	Degrees of freedom	Sum of squares	Mean square	f
Mean	1	648.2700		
Levels of A	2	12.7400	6.3700	25.82
Levels of B	1	0.4033	0.4033	1.63
Interactions	2	1.2067	0.6033	2.45
Residual	6	1.4800	0.2467	
Total	12	664.1		

Another feature of this analysis is that the hypothesis tests are independent in the sense that the results are not affected by which terms – other than those relating to the hypothesis in question – are also included in the model. For example, the hypothesis of no differences due to factor B, $H_B : \beta_k = 0$ for all k, could equally well be tested using either (1) models $E(Y_{jkl}) = \mu + \alpha_j + \beta_k$ and $E(Y_{jkl}) = \mu + \alpha_j$ and hence

$$\sigma^2 D_B - \sigma^2 D_I = 3.0900 - 2.6867 = 0.4033$$

or (2) models $E(Y_{jkl}) = \mu + \beta_k$ and $E(Y_{jkl}) = \mu$ and hence

$$\sigma^2 D_M - \sigma^2 D_A = 15.8300 - 15.4267 = 0.4033$$

The reason for this feature is that the data are **balanced**, that is, there are equal numbers of observations in each subclass. As a result it is possible to specify the design matrix X in such a way that there are orthogonal components corresponding to each of the models of interest and so the hypothesis tests are independent (see section 6.9). Details of an orthogonal parametrization for the study design illustrated by these data are given at the end of Appendix C. An example in which the hypothesis tests are not independent is given in Exercise 7.3.

The conventional ANOVA summary of the hypothesis tests for this data set is shown in Table 7.6. The first number in the 'sum of squares' column is the value of $\mathbf{b}^T X^T \mathbf{y}$ corresponding to the model $E(Y_{jkl}) = \mu$.

The second is the difference in values of $\mathbf{b}^T X^T \mathbf{y}$ for the models

$$E(Y_{jkl}) = \mu + \alpha_j + \beta_k \quad \text{and} \quad E(Y_{jkl}) = \mu + \beta_k$$

and similarly for the third number. The fourth is the difference in values of $\mathbf{b}^T X^T \mathbf{y}$ for the saturated model and the additive model. The 'residual sum of squares' is $\sigma^2 D_S$, i.e. the 'total sum of squares', $\mathbf{y}^T\mathbf{y}$, minus the value of $\mathbf{b}^T X^T \mathbf{y}$ for the saturated model. The degrees of freedom are obtained analogously.

7.5 CROSSED AND NESTED FACTORS

In the example in section 7.4 the factors A and B are said to be **crossed** because there is a subclass corresponding to each combination of levels A_j and B_k and all the comparisons represented by the terms α_j and β_k and $(\alpha\beta)_{jk}$ in the saturated model

$$E(Y_{jkl}) = \mu + \alpha_j + \beta_k + (\alpha\beta)_{jk}, \qquad j = 1, \ldots, J,\ k = 1, \ldots, K$$

are of potential interest.

This contrasts with the two-factor **nested** design shown in Table 7.7 which represents an experiment to compare two drugs (A_1 and A_2) one of which is tested in three hospitals (B_1, B_2 and B_3) and the other in two hospitals (B_4 and B_5). We want to compare the effects of the two drugs and possible differences in response among hospitals using the same drug. It is not sensible to make comparisons among hospitals which use different drugs. The saturated model involves the parameters μ, α_1, α_2, $(\alpha\beta)_{11}$, $(\alpha\beta)_{12}$, $(\alpha\beta)_{13}$, $(\alpha\beta)_{24}$ and $(\alpha\beta)_{25}$. The conventional sum-to-zero constraints are $\alpha_1 + \alpha_2 = 0$, $(\alpha\beta)_{11} + (\alpha\beta)_{12} + (\alpha\beta)_{13} = 0$ and $(\alpha\beta)_{24} + (\alpha\beta)_{25} = 0$, or the corner-point constraints are $\alpha_1 = 0$, $(\alpha\beta)_{11} = 0$ and $(\alpha\beta)_{24} = 0$. To test the hypothesis of no difference between drugs (but allowing for differences among hospitals using the same drug) the saturated model is compared to a model with parameters μ, β_1, β_2, β_3, β_4 and β_5, where the β's are the hospital effects and are subject to the constraints $\beta_1 + \beta_2 + \beta_3 = 0$ and $\beta_4 + \beta_5 = 0$, or $\beta_1 = 0$ and $\beta_4 = 0$.

Table 7.7 Nested two-factor experiment

	Drug A_1			Drug A_2	
Hospitals	B_1	B_2	B_3	B_4	B_5
Responses	Y_{111}	Y_{121}	Y_{131}	Y_{241}	Y_{251}
	$\vdots$	$\vdots$	$\vdots$	$\vdots$	$\vdots$
	Y_{11n_1}	Y_{12n_2}	Y_{13n_3}	Y_{24n_4}	Y_{25n_5}

7.6 MORE COMPLICATED MODELS

Analysis of variance models can readily be defined for more complicated hypotheses and for study designs with more than two factors. The factors may be crossed or nested or some mixture of these forms. The models can include higher-order interaction terms such as $(\alpha\beta\gamma)_{jkl}$ as well as the first-order interactions like $(\alpha\beta)_{jk}$ and the main effects. These extensions do not involve any fundamental differences from the examples already considered so they are not examined further in this book.

In all the above examples we considered only hypotheses in which certain parameters in the saturated model are omitted in the reduced models. For instance, in the plant weight example (Table 7.1) the saturated model $E(Y_{jk}) = \mu + \alpha_j$ was compared with the reduced model $E(Y_{jk}) = \mu$ corresponding to the hypothesis that $\alpha_1 = \alpha_2 = \alpha_3 = 0$. Sometimes we are interested in testing more complicated hypotheses, for example that the control condition and treatment A in the plant weight experiment are equally effective but different from the treatment B, i.e. $\alpha_1 = \alpha_2$ but α_3 is not necessarily the same. Such hypotheses can be readily accommodated in the model-fitting approach by the appropriate choice of parameters and dummy variables, for example the hypothesis $\alpha_1 = \alpha_2$ is equivalent to fitting $E(Y_{1k}) = E(Y_{2k}) = \beta_1$ and $E(Y_{3k}) = \beta_2$.

In general, multiple hypothesis tests are not independent. The only exception is if there is a design matrix with orthogonal components so that the total sum of squares can be partitioned into disjoint terms corresponding to the hypotheses (as illustrated in Table 7.6). Usually this is only possible if the hypotheses are particularly simple (e.g. the interaction and main effects are zero) and if the experimental design is balanced (i.e. there are equal numbers of observations in each subclass). If the hypotheses are not independent then care is needed in interpreting simultaneous significance tests.

7.7 CHOICE OF CONSTRAINT EQUATIONS AND DUMMY VARIABLES

The numerical examples considered above illustrate several major issues relating to the choice of constraint equations and dummy variables for ANOVA models.

ANOVA models are usually specified in terms of parameters which are readily interpretable as effects due to factor levels and interactions. However, the number of parameters is usually larger than the number of independent normal equations. Therefore extra equations, traditionally in the form of sum-to-zero constraints, are added. (If the design is unbalanced there is some controversy about the most appropriate choice of constraint equations.) In the framework of generalized linear models this means that the equations (7.2) are not the normal equations obtained by the methods of maximum likelihood or least squares. Therefore the standard computational procedures cannot be used. Also the terms of $\boldsymbol{\beta}$ are generally not identifiable, and unique unbiased point estimates and confidence intervals can only be obtained for certain linear combinations of parameters, called **estimable functions**. Nevertheless, if the main purpose of analysing the data is to test hypotheses, the use of sum-to-zero constraints is entirely appropriate and convenient provided that special purpose computer programs are used. Most of the major statistical computing packages use this method.

If the corner-point constraints are used the elements of $\boldsymbol{\beta}$ and the corresponding columns of $\boldsymbol{X}$ are arranged as

$$\boldsymbol{\beta} = \begin{bmatrix} \boldsymbol{\beta}_1 \\ \boldsymbol{\beta}_2 \end{bmatrix}$$

and $\boldsymbol{X} = [\boldsymbol{X}_1, \boldsymbol{X}_2]$ so that $\boldsymbol{X}_1^{\mathrm{T}}\boldsymbol{X}_1$ is non-singular and $\boldsymbol{\beta}_2$ is set to $\mathbf{0}$. Thus

$$E(\mathbf{y}) = \boldsymbol{X\beta} = \boldsymbol{X}_1\boldsymbol{\beta}_1$$

Then the normal equations

$$\boldsymbol{X}_1^{\mathrm{T}}\boldsymbol{X}_1\mathbf{b}_1 = \boldsymbol{X}_1^{\mathrm{T}}\mathbf{y}$$

can be solved using standard multiple regression or generalized linear modelling programs and the estimators have various desirable properties (e.g. $\mathbf{b}_1$ is unbiased and has variance–covariance matrix $\sigma^2(\boldsymbol{X}_1^{\mathrm{T}}\boldsymbol{X}_1)^{-1}$). However, the interpretation of parameters subject to corner-point constraints is perhaps less straightforward than with sum-to-zero constraints. Also all the calculations usually have to be repeated for each new model fitted. In practice, estimation using corner-point constraints is performed so that parameters are estimated sequentially in such a way that the redundant corner-point parameters (which are said to be **aliased**) are systematically identified and set equal to zero (e.g. this is the procedure used in GLIM).

In the two-factor ANOVA example in section 7.4, the most elegant analysis was obtained by choosing the dummy variables so that the design matrix $\boldsymbol{X}$ had orthogonal components corresponding to each of the hypotheses to be tested. For simple well-planned experiments where this form of analysis is possible there are computational benefits (e.g. parameter estimates are the same for all models) and advantages in interpretation (e.g. independence of the hypothesis tests). However, for unbalanced experimental designs or hypotheses involving more complicated contrasts, it is unlikely that orthogonal forms exist.

In summary, for any particular sequence of models the choice of constraints and dummy variables affects the computational procedures and the parameter estimates. Provided the same models are used, it does not, however, influence the results of hypothesis tests. The reason is that any solution $\mathbf{b}$ of the normal equations (7.1) corresponds to the unique minimum of $(\mathbf{y} - X\boldsymbol{\beta})^{\mathrm{T}}(\mathbf{y} - X\boldsymbol{\beta})$. Hence the statistics $\sigma^2 D = \mathbf{y}^{\mathrm{T}}\mathbf{y} - \mathbf{b}^{\mathrm{T}}X^{\mathrm{T}}\mathbf{y}$ are the same regardless of the way the models are specified.

7.8 ANALYSIS OF COVARIANCE

This is the term used for mixed models in which some of the explanatory variables are dummy variables representing factor levels and others are continuous measurements, called covariates. As with ANOVA we are interested in comparing means of subclasses defined by factor levels but, recognizing that the covariates may also affect the responses, we compare the means after 'adjustment' for covariate effects.

A typical example is provided by the data in Table 7.8. The responses Y_{jk} are achievement scores measured at three levels of a factor representing three different training methods, and the covariates x_{jk} are aptitude scores measured before training commenced. We want to compare the training methods, taking into account differences in initial aptitude between the three groups of subjects.

The data are plotted in Fig. 7.1. There is evidence that the achievement scores y increase linearly with aptitude x and that the y values are generally higher for treatment groups A_2 and A_3 than for A_1.

Table 7.8 Achievement scores (data from Winer, 1971, p. 776)

Training method	A_1		A_2		A_3	
	y	x	y	x	y	x
	6	3	8	4	6	3
	4	1	9	5	7	2
	5	3	7	5	7	2
	3	1	9	4	7	3
	4	2	8	3	8	4
	3	1	5	1	5	1
	6	4	7	2	7	4
Total	31	15	53	24	47	19
Sum of squares	147	41	413	96	321	59
$\sum xy$	75		191		132	

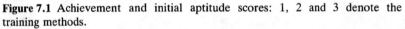

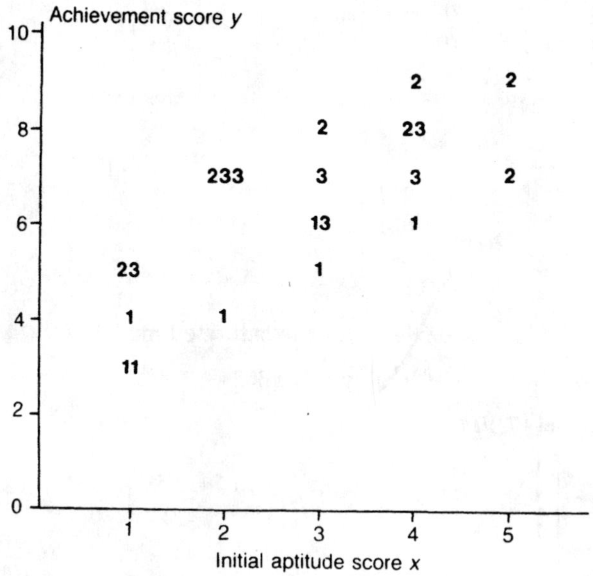

Figure 7.1 Achievement and initial aptitude scores: 1, 2 and 3 denote the training methods.

To test the hypothesis that there are no differences in mean achievement scores among the three training methods, after adjustment for initial aptitude, we compare the saturated model

$$E(Y_{jk}) = \mu_j + \gamma x_{jk} \tag{7.9}$$

with the reduced model

$$E(Y_{jk}) = \mu + \gamma x_{jk} \tag{7.10}$$

where $j = 1, 2, 3$ and $k = 1, \ldots, 7$. Let

$$\mathbf{y}_j = \begin{bmatrix} Y_{j1} \\ \vdots \\ Y_{j7} \end{bmatrix} \quad \text{and} \quad \mathbf{x}_j = \begin{bmatrix} x_{j1} \\ \vdots \\ x_{j7} \end{bmatrix}$$

so that in matrix notation the saturated model (7.9) is $E(\mathbf{y}) = X\boldsymbol{\beta}$ with

$$\mathbf{y} = \begin{bmatrix} \mathbf{y}_1 \\ \mathbf{y}_2 \\ \mathbf{y}_3 \end{bmatrix} \quad \boldsymbol{\beta} = \begin{bmatrix} \mu_1 \\ \mu_2 \\ \mu_3 \\ \gamma \end{bmatrix} \quad \text{and} \quad X = \begin{bmatrix} \mathbf{1} & \mathbf{0} & \mathbf{0} & \mathbf{x}_1 \\ \mathbf{0} & \mathbf{1} & \mathbf{0} & \mathbf{x}_2 \\ \mathbf{0} & \mathbf{0} & \mathbf{1} & \mathbf{x}_3 \end{bmatrix}$$

where **0** and **1** are vectors of length 7. Then

$$X^TX = \begin{bmatrix} 7 & 0 & 0 & 15 \\ 0 & 7 & 0 & 24 \\ 0 & 0 & 7 & 19 \\ 15 & 24 & 19 & 196 \end{bmatrix} \qquad X^T\mathbf{y} = \begin{bmatrix} 31 \\ 53 \\ 47 \\ 398 \end{bmatrix}$$

and so

$$\mathbf{b} = \begin{bmatrix} 2.837 \\ 5.024 \\ 4.698 \\ 0.743 \end{bmatrix}$$

Also $\mathbf{y}^T\mathbf{y} = 881$ and $\mathbf{b}^T X^T \mathbf{y} = 870.698$ so for the saturated model (7.9)

$$\sigma^2 D_1 = \mathbf{y}^T\mathbf{y} - \mathbf{b}^T X^T \mathbf{y} = 10.302$$

For the reduced model (7.10)

$$\boldsymbol{\beta} = \begin{bmatrix} \mu \\ \gamma \end{bmatrix} \qquad X = \begin{bmatrix} \mathbf{1} & \mathbf{x}_1 \\ \mathbf{1} & \mathbf{x}_2 \\ \mathbf{1} & \mathbf{x}_3 \end{bmatrix} \quad \text{so} \quad X^TX = \begin{bmatrix} 21 & 58 \\ 58 & 196 \end{bmatrix}$$

and

$$X^T\mathbf{y} = \begin{bmatrix} 131 \\ 398 \end{bmatrix}$$

Hence

$$\mathbf{b} = \begin{bmatrix} 3.447 \\ 1.011 \end{bmatrix} \qquad \mathbf{b}^T X^T \mathbf{y} = 853.766 \quad \text{and so} \quad \sigma^2 D_0 = 27.234$$

If we assume that the saturated model (7.9) is correct, then $D_1 \sim \chi^2_{17}$. If the null hypothesis corresponding to model (7.10) is true then $D_0 \sim \chi^2_{19}$ so

$$f = \frac{D_0 - D_1}{2\sigma^2} \bigg/ \frac{D_1}{17\sigma^2} \sim F_{2,17}$$

Table 7.9 ANCOVA table for data in Table 7.8

Source of variation	Degrees of freedom	Sum of squares	Mean square	f
Mean and covariate	2	853.766		
Factor levels	2	16.932	8.466	13.97
Residual	17	10.302	0.606	
Total	21	881.000		

For these data

$$f = \frac{16.932}{2} \Big/ \frac{10.302}{17} = 13.97$$

indicating a significant difference in achievement scores for the training methods, after adjustment for initial differences in aptitude. The usual presentation of this analysis is given in Table 7.9; the values are obtained as explained in section 7.4.

7.9 EXERCISES

7.1 Table 7.10 shows plasma inorganic phosphate levels (mg/dl) one hour after a standard glucose tolerance test for hyperinsulinemic and non-hyperinsulinemic obese subjects and controls (data from Jones, 1987).

Table 7.10

Hyperinsulinemic obese	Non-hyperinsulinemic obese	Controls
2.3	3.0	3.0
4.1	4.1	2.6
4.2	3.9	3.1
4.0	3.1	2.2
4.6	3.3	2.1
4.6	2.9	2.4
3.8	3.3	2.8
5.2	3.9	3.4
3.1		2.9
3.7		2.6
3.8		3.1
		3.2

(a) Perform a one-factor analysis of variance to test the hypothesis that there are no mean differences among the three groups. What conclusions can you draw?

(b) Obtain a 95% confidence interval for the difference in means between the two obese groups.

(c) Using an appropriate model examine the standardized residuals for all the observations to look for any systematic effects and to check the Normality assumption.

7.2 The weight (in grams) of machine components of a standard size made by four different workers on two different days are shown in Table 7.11; five components were chosen randomly from the output of each worker on each day. Perform a complete analysis of variance testing for differences among workers, between days and possible interaction effects. What are your conclusions?

Table 7.11

	Workers			
	1	2	3	4
Day 1	35.7	38.4	34.9	37.1
	37.1	37.2	34.3	35.5
	36.7	38.1	34.5	36.5
	37.7	36.9	33.7	36.0
	35.3	37.2	36.2	33.8
Day 2	34.7	36.9	32.0	35.8
	35.2	38.5	35.2	32.9
	34.6	36.4	33.5	35.7
	36.4	37.8	32.9	38.0
	35.2	36.1	33.3	36.1

7.3 Table 7.12 shows data from a fictitious two-factor experiment.

(a) Test the hypothesis that there are no interaction effects.
(b) Test the hypothesis that there is no effect due to factor A
   (i) by comparing the models

$$E(Y_{jkl}) = \mu + \alpha_j + \beta_k \quad \text{and} \quad E(Y_{jkl}) = \mu + \beta_k$$

   (ii) by comparing the models

$$E(Y_{jkl}) = \mu + \alpha_j \quad \text{and} \quad E(Y_{jkl}) = \mu$$

Explain the results.

Table 7.12

	Factor B	
Factor A	B_1	B_2
A_1	5	3, 4
A_2	6, 4	4, 3
A_3	7	6, 8

7.4 For the achievement score data in Table 7.8:

(a) Test the hypothesis that the treatment effects are equal, ignoring the covariate, i.e. using the models $E(Y_{jk}) = \mu_j$ and $E(Y_{jk}) = \mu$, and compare your results with Table 7.9.

(b) Test the hypothesis that the initial aptitude has the same effect for all training methods by comparing the models

$$E(Y_{jk}) = \mu_j + \alpha_j x_{jk} \quad \text{and} \quad E(Y_{jk}) = \mu_j + \alpha x_{jk}$$

(this is possible with GLIM but some other programs do not allow covariates to have group-specific slope parameters).

7.5 Show that the minimum value of $(\mathbf{y} - X\boldsymbol{\beta})^{\mathrm{T}}(\mathbf{y} - X\boldsymbol{\beta})$ is given by any solution $\boldsymbol{\beta} = \mathbf{b}$ of the Normal equations $X^{\mathrm{T}}X\mathbf{b} = X^{\mathrm{T}}\mathbf{y}$. If $\mathbf{b}_1$ and $\mathbf{b}_2$ are two such solutions it cannot be that

$$(\mathbf{y} - X\mathbf{b}_1)^{\mathrm{T}}(\mathbf{y} - X\mathbf{b}_1) > (\mathbf{y} - X\mathbf{b}_2)^{\mathrm{T}}(\mathbf{y} - X\mathbf{b}_2)$$

or

$$(\mathbf{y} - X\mathbf{b}_1)^{\mathrm{T}}(\mathbf{y} - X\mathbf{b}_1) < (\mathbf{y} - X\mathbf{b}_2)^{\mathrm{T}}(\mathbf{y} - X\mathbf{b}_2)$$

Hence $(\mathbf{y} - X\mathbf{b}_1)^{\mathrm{T}}(\mathbf{y} - X\mathbf{b}_1)$ must equal $(\mathbf{y} - X\mathbf{b}_2)^{\mathrm{T}}(\mathbf{y} - X\mathbf{b}_2)$ and so $\mathbf{b}_1^{\mathrm{T}}X^{\mathrm{T}}\mathbf{y} = \mathbf{b}_2^{\mathrm{T}}X^{\mathrm{T}}\mathbf{y}$.

8

Binary variables and logistic regression

8.1 PROBABILITY DISTRIBUTIONS

In this chapter we consider generalized linear models in which the outcome variables are measured on a binary scale. For example, the responses may be alive or dead, or present or absent. 'Success' and 'failure' are used as generic terms for the two categories.

First we define the binary random variable

$$Z = \begin{cases} 1 \text{ if the outcome is a success} \\ 0 \text{ if the outcome is a failure} \end{cases}$$

with $\Pr(Z = 1) = \pi$ and $\Pr(Z = 0) = 1 - \pi$. If there are n such random variables $Z_1, \ldots, Z_n$ which are independent with $\Pr(Z_j = 1) = \pi_j$, then their joint probability is

$$\prod_{j=1}^{n} \pi_j^{z_j}(1 - \pi_j)^{1-z_j} = \exp\left[\sum_{j=1}^{n} z_j \log\left(\frac{\pi_j}{1 - \pi_j}\right) + \sum_{j=1}^{n} \log(1 - \pi_j)\right] \quad (8.1)$$

which is a member of the exponential family (see equation (3.6)).

Next for the case where the π_j's are all equal, we can define

$$Y = \sum_{j=1}^{n} Z_j$$

so that Y is the number of successes in n 'trials'. The random variable Y has the binomial distribution $b(n, \pi)$:

$$\Pr(Y = y) = \binom{n}{y}\pi^y(1 - \pi)^{n-y} \qquad y = 0, 1, \ldots, n \qquad (8.2)$$

Finally we consider the general case of N independent random variables $Y_1, Y_2, \ldots, Y_N$ corresponding to the numbers of successes in N different subgroups or strata (Table 8.1). If $Y_i \sim b(n_i, \pi_i)$ the log-likelihood function is

$$l(\pi_1, \ldots, \pi_N; y_1, \ldots, y_N)$$
$$= \sum_{i=1}^{N} \left[y_i \log\left(\frac{\pi_i}{1-\pi_i}\right) + n_i \log(1-\pi_i) + \log\binom{n_i}{y_i} \right] \quad (8.3)$$

The distribution (8.3) does not correspond directly to equation (3.6) for the exponential family because the n_i's may not all be the same. Nevertheless if the joint distribution of the Y_i's is written in terms of the binary variables Z_j it follows from (8.1) that (8.3) does belong to the exponential family of distributions.

Table 8.1 Frequencies for N binomial distributions

	Subgroups			
	1	2	...	N
Successes	Y_1	Y_2	...	Y_N
Failures	$n_1 - Y_1$	$n_2 - Y_2$	...	$n_N - Y_N$
Totals	n_1	n_2	...	n_N

8.2 GENERALIZED LINEAR MODELS

We want to describe the proportion of successes, $P_i = Y_i/n_i$, in each subgroup in terms of factor levels and other explanatory variables which characterize the subgroup. We do this by modelling the probabilities π_i as

$$g(\pi_i) = \mathbf{x}_i^{\mathrm{T}}\boldsymbol{\beta}$$

where $\mathbf{x}_i$ is a vector of explanatory variables (dummy variables for factor levels and measured values of covariates), $\boldsymbol{\beta}$ is a vector of parameters and g is a link function.

The simplest case is the **linear model**

$$\pi = \mathbf{x}^{\mathrm{T}}\boldsymbol{\beta}$$

This is used in some practical applications but it has the disadvantage that although π is a probability the fitted values $\mathbf{x}^{\mathrm{T}}\mathbf{b}$ may be outside the interval [0, 1].

To ensure that π is restricted to the interval [0, 1] we often model it using a cumulative probability distribution

$$\pi = g^{-1}(\mathbf{x}^{\mathrm{T}}\boldsymbol{\beta}) = \int_{-\infty}^{t} f(s)\,\mathrm{d}s$$

where $f(s) \geqslant 0$ and $\int_{-\infty}^{\infty} f(s)\,\mathrm{d}s = 1$. The probability density function $f(s)$

is called the **tolerance distribution**. Some commonly used examples are considered in section 8.3.

8.3 DOSE RESPONSE MODELS

Historically one of the first uses of regression-like models for binomial data was for bioassay results (Finney, 1973). Responses were the proportions or percentages of 'successes', for example, the proportion of experimental animals killed by various dose levels of a toxic substance. Such data are sometimes called **quantal responses**. The aim is to describe the probability of 'success', π, as a function of the dose, x, for example $g(\pi) = \beta_1 + \beta_2 x$.

If the tolerance distribution $f(s)$ is the uniform distribution on the interval $[c_1, c_2]$

$$f(s) = \begin{cases} \dfrac{1}{c_2 - c_1} & \text{if } c_1 \leqslant s \leqslant c_2 \\ 0 & \text{otherwise} \end{cases}$$

then

$$\pi = \int_{c_1}^{x} f(s)\,\mathrm{d}s = \frac{x - c_1}{c_2 - c_1} \qquad \text{for } c_1 \leqslant x \leqslant c_2$$

(see Fig. 8.1). This is of the form

$$\pi = \beta_1 + \beta_2 x \qquad \text{where } \beta_1 = \frac{-c_1}{c_2 - c_1} \quad \text{and} \quad \beta_2 = \frac{1}{c_2 - c_1}$$

Figure 8.1 Uniform distribution on $[c_1, c_2]$.

This **linear model** is equivalent to using the identity function as the link function g and imposing conditions on x, β_1 and β_2 corresponding to $c_1 \leqslant x \leqslant c_2$. These extra conditions mean that the standard methods for estimating β_1 and β_2 for generalized linear models cannot be directly applied. In practice this model is not widely used.

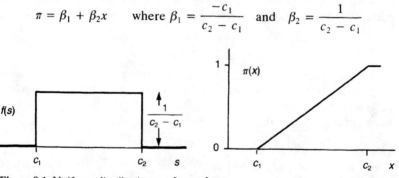

One of the original models used for bioassay data is called the **probit model**. The normal distribution is used as the tolerance distribution (see Fig. 8.2).

$$\pi = \frac{1}{\sigma\sqrt{2\pi}} \int_{-\infty}^{x} \exp\left[-\frac{1}{2}\left(\frac{s-\mu}{\sigma}\right)^2\right] ds$$

$$= \Phi\left(\frac{x-\mu}{\sigma}\right)$$

where Φ denotes the cumulative probability function for the standard Normal distribution $N(0, 1)$. Thus

$$\Phi^{-1}(\pi) = \beta_1 + \beta_2 x$$

where $\beta_1 = -\mu/\sigma$ and $\beta_2 = 1/\sigma$ and the link function g is the inverse cumulative Normal probability function Φ^{-1}. Probit models are used in several areas of biological and social sciences in which there are natural interpretations of the model; for example, $x = \mu$ is called the **median lethal dose** LD(50) because it corresponds to the dose required to kill half the animals, on the average.

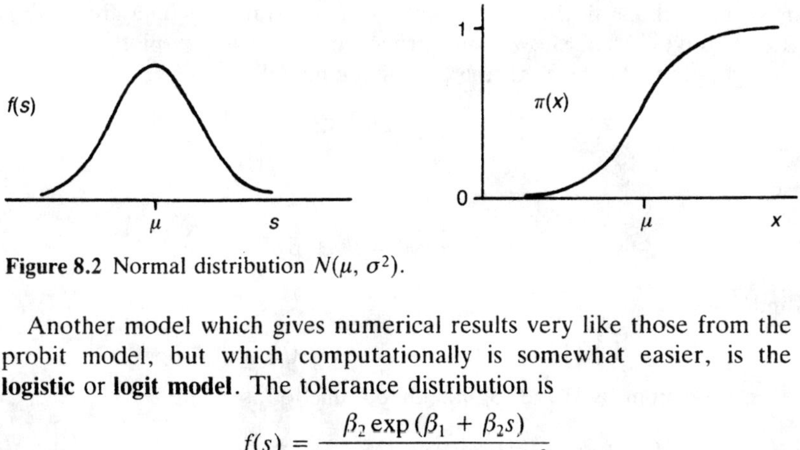

Figure 8.2 Normal distribution $N(\mu, \sigma^2)$.

Another model which gives numerical results very like those from the probit model, but which computationally is somewhat easier, is the **logistic** or **logit model**. The tolerance distribution is

$$f(s) = \frac{\beta_2 \exp(\beta_1 + \beta_2 s)}{[1 + \exp(\beta_1 + \beta_2 s)]^2}$$

so

$$\pi = \int_{-\infty}^{x} f(s)\, ds = \frac{\exp(\beta_1 + \beta_2 x)}{1 + \exp(\beta_1 + \beta_2 x)}$$

This gives the link function as

$$\log\left(\frac{\pi}{1-\pi}\right) = \beta_1 + \beta_2 x$$

$\log[\pi/(1-\pi)]$ is sometimes called the **logit function** and it has a natural interpretation as the logarithm of odds (see Exercise 8.2). The logistic model is widely used for binomial data and is implemented in many statistical programs. The shapes of the functions $f(s)$ and $\pi(x)$ are similar to those for the probit model (Fig. 8.2) except in the tails of the distributions (see Cox and Snell, 1989).

Several other models are also used for dose response data. For example, if the **extreme value distribution**

$$f(s) = \beta_2 \exp[(\beta_1 + \beta_2 s) - \exp(\beta_1 + \beta_2 s)]$$

is used as the tolerance distribution then

$$\pi = 1 - \exp[-\exp(\beta_1 + \beta_2 x)]$$

and so $\log[-\log(1-\pi)] = \beta_1 + \beta_2 x$. This link, $\log[-\log(1-\pi)]$, is called the **complementary log log function**. The model is similar to the logistic and probit models for values of π near 0.5 but differs from them for π near 0 or 1. These models are illustrated in the following example.

Example 8.1 Dose–response models

Table 8.2 shows numbers of insects dead after five hours' exposure to gaseous carbon disulphide at various concentrations (data from Bliss, 1935). Figure 8.3 shows the proportions $p_i = y_i/n_i$ plotted against dosage x_i. We begin by fitting the logistic model

$$\pi_i = \frac{\exp(\beta_1 + \beta_2 x_i)}{1 + \exp(\beta_1 + \beta_2 x_i)}$$

so

$$\log\left(\frac{\pi_i}{1-\pi_i}\right) = \beta_1 + \beta_2 x_i$$

and

$$\log(1-\pi_i) = -\log[1 + \exp(\beta_1 + \beta_2 x_i)]$$

Therefore from (8.3) the log-likelihood function is

$$l = \sum_{i=1}^{N}\left[y_i(\beta_1 + \beta_2 x_i) - n_i \log[1 + \exp(\beta_1 + \beta_2 x_i)] + \log\binom{n_i}{y_i}\right]$$

and the scores with respect to β_1 and β_2 are

$$U_1 = \frac{\partial l}{\partial \beta_1} = \sum\left\{y_i - n_i\left[\frac{\exp(\beta_1 + \beta_2 x_i)}{1 + \exp(\beta_1 + \beta_2 x_i)}\right]\right\} = \sum(y_i - n_i\pi_i)$$

$$U_2 = \frac{\partial l}{\partial \beta_2} = \sum\left\{y_i x_i - n_i x_i\left[\frac{\exp(\beta_1 + \beta_2 x_i)}{1 + \exp(\beta_1 + \beta_2 x_i)}\right]\right\} = \sum x_i(y_i - n_i\pi_i)$$

Table 8.2 Beetle mortality data

Dose x_i ($\log_{10} CS_2$ mg l^{-1})	Number of insects, n_i	Number killed, y_i
1.6907	59	6
1.7242	60	13
1.7552	62	18
1.7842	56	28
1.8113	63	52
1.8369	59	53
1.8610	62	61
1.8839	60	60

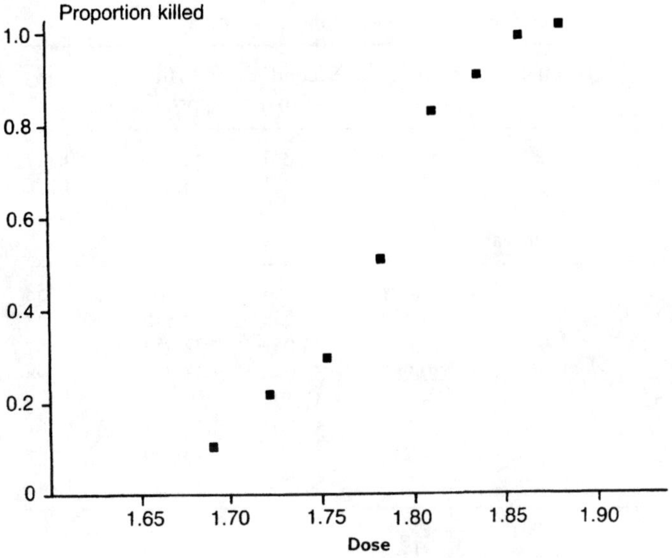

Figure 8.3 Beetle mortality data.

Similarly the information matrix is

$$\mathcal{J} = \begin{bmatrix} \sum n_i\pi_i(1-\pi_i) & \sum n_i x_i \pi_i(1-\pi_i) \\ \sum n_i x_i \pi_i(1-\pi_i) & \sum n_i x_i^2 \pi_i(1-\pi_i) \end{bmatrix}$$

Maximum likelihood estimates are obtained by solving the iterative equation

$$\mathcal{J}^{(m-1)}\mathbf{b}^{(m)} = \mathcal{J}^{(m-1)}\mathbf{b}^{(m-1)} + U^{(m-1)}$$

(from (4.7)) where the superscript (m) indicates the mth approximation and $\mathbf{b}$ is the vector of estimates. Starting from $b_1^{(0)} = 0$ and $b_2^{(0)} = 0$ successive approximations are shown in Table 8.3 together with the fitted values $\hat{y}_i = n_i\hat{\pi}_i$. The estimated variance–covariance matrix for $\mathbf{b}$ is $[\mathcal{J}(\mathbf{b})]^{-1}$ (from section 5.3). The log-likelihood ratio statistic is

$$D = \sum_{i=1}^{N}\left[y_i \log\left(\frac{y_i}{\hat{y}_i}\right) + (n_i - y_i)\log\left(\frac{n - y_i}{n - \hat{y}_i}\right)\right]$$

where $\hat{y}_i$ denotes the fitted value (see Exercise 5.2(a), also section 8.5).

Table 8.3 Fitting the logistic model to the beetle mortality data

		Initial estimate	First approx.	Second approx.	Fourth approx.	Tenth approx.
	b_1	0	−37.849	−53.851	−60.700	−60.717
	b_2	0	21.334	30.382	34.261	34.270
Observations		Fitted values				
y_1	6	29.5	8.508	4.544	3.460	3.458
y_2	13	30.0	15.369	11.254	9.845	9.842
y_3	18	31.0	24.810	23.059	22.454	22.451
y_4	28	28.0	30.983	32.946	33.896	33.898
y_5	52	31.5	43.361	48.197	50.092	50.096
y_6	53	29.5	46.739	51.704	53.288	53.291
y_7	61	31.0	53.593	58.060	59.220	59.222
y_8	60	30.0	54.732	58.036	58.742	58.743

$$[\mathcal{J}(\mathbf{b})]^{-1} = \begin{bmatrix} 26.802 & 15.061 \\ 15.061 & 8.469 \end{bmatrix}, \quad D = 11.23$$

The estimates are $b_1 = -60.72$ and $b_2 = 34.27$ and their standard errors are $\sqrt{26.802} = 5.18$ and $\sqrt{8.469} = 2.91$ respectively. If the logistic model provides a good summary of the data, the log-likelihood ratio statistic D has an approximate χ^2_6 distribution because there are $N = 8$ observations and $p = 2$ parameters. But the upper 5% point of the χ^2_6 distribution is 12.59 which suggests that the model does not fit the data particularly well.

Using the program GLIM several alternative models were fitted to these data:

1. Logistic (with the logit link function);
2. Probit (with the inverse cumulative Normal link function Φ^{-1});
3. Extreme value (with the complementary log log link function).

The results are shown in Table 8.4. Among these models the extreme value model clearly provides the best description of the data.

Table 8.4 Comparison of various dose–response models for the beetle mortality data

Observed value of Y	Logistic model	Probit model	Extreme value model
6	3.46	3.36	5.59
13	9.84	10.72	11.28
18	22.45	23.48	20.95
28	33.90	33.82	30.37
52	50.10	49.62	47.78
53	53.29	53.32	54.14
61	59.22	59.66	61.11
60	58.74	59.23	59.95
D	11.23	10.12	3.45

8.4 GENERAL LOGISTIC REGRESSION

The simple logistic model $\log[\pi_i/(1-\pi_i)] = \beta_1 + \beta_2 x_i$ used in example 8.1 is a special case of the general logistic regression model

$$\text{logit}\,\pi_i = \log\left(\frac{\pi_i}{1-\pi_i}\right) = \mathbf{x}_i^{\mathrm{T}}\boldsymbol{\beta}$$

where $\mathbf{x}_i$ is a vector of continuous measurements corresponding to covariates and dummy variables corresponding to factor levels and $\boldsymbol{\beta}$ is the parameter vector. This model is very widely used for analysing multivariate data involving binary responses. It provides a powerful technique analogous to multiple regression and ANOVA for continuous responses. Computer programs for performing logistic regression are available in most statistical packages, for example, the program PLR in BMDP or the procedure PROC LOGIST in SAS.

8.5 MAXIMUM LIKELIHOOD ESTIMATION AND THE LOG-LIKELIHOOD RATIO STATISTIC

For any of the dose–response models and for extensions such as the general logistic model maximum likelihood estimates of the parameters $\boldsymbol{\beta}$ and consequently of the probabilities $\pi_i = g^{-1}(\mathbf{x}_i^{\mathrm{T}}\boldsymbol{\beta})$, are obtained by maximizing the log-likelihood function

$$l(\boldsymbol{\pi}; \mathbf{y}) = \sum_{i=1}^{N} \left[y_i \log \pi_i + (n_i - y_i) \log (1 - \pi_i) + \log \binom{n_i}{y_i} \right]$$

using the methods described in Chapter 4. Maximum likelihood estimation is possible even if $n_i = 1$ and/or $y_i = 0$ (unlike some of the least squares methods described in section 8.7).

To measure the goodness of fit of a model we use the log-likelihood ratio statistic

$$D = 2[l(\hat{\boldsymbol{\pi}}_{\max}; \mathbf{y}) - l(\hat{\boldsymbol{\pi}}; \mathbf{y})]$$

where $\hat{\boldsymbol{\pi}}_{\max}$ is the vector of maximum likelihood estimates corresponding to the maximal model and $\hat{\boldsymbol{\pi}}$ is the vector of maximum likelihood estimates for the model of interest.

Without loss of generality, for the maximal model we take the π_i's as the parameters to be estimated. Then

$$\frac{\partial l}{\partial \pi_i} = \frac{y_i}{\pi_i} - \frac{n_i - y_i}{1 - \pi_i}$$

so the ith element of $\hat{\boldsymbol{\pi}}_{\max}$, the solution of the equation $\partial l / \partial \pi_i = 0$, is y_i / n_i (i.e. the observed proportion of successes in the ith subgroup). Hence

$$l(\hat{\boldsymbol{\pi}}_{\max}; \mathbf{y}) = \sum_{i=1}^{N} \left[y_i \log \left(\frac{y_i}{n_i} \right) + (n_i - y_i) \log \left(1 - \frac{y_i}{n_i} \right) + \log \binom{n_i}{y_i} \right]$$

and so

$$D = 2 \sum_{i=1}^{N} \left[y_i \log \left(\frac{y_i}{n_i \hat{\pi}_i} \right) + (n_i - y_i) \log \left(\frac{n_i - y_i}{n_i - n_i \hat{\pi}_i} \right) \right] \qquad (8.4)$$

Thus D has the form

$$D = 2 \sum o \log \frac{o}{e}$$

where o denotes the observed frequencies y_i and $(n_i - y_i)$ from the cells of Table 8.1 and e denotes the corresponding estimated expected frequencies or fitted values $n_i \hat{\pi}_i$ and $(n_i - n_i \hat{\pi}_i)$. Summation is over all $2 \times N$ cells of the table.

Notice that D does not involve any nuisance parameters (unlike σ^2

for Normal response data), and so goodness of fit can be assessed and hypotheses can be tested directly using the approximation

$$D \sim \chi^2_{N-p}$$

where p is the number of β parameters estimated.

Example 8.2 Use of generalized logistic regression models
The data (Table 8.5), cited by Wood (1978) are taken from Sangwan-Norrell (1977). They are numbers y_{jk} of embryogenic anthers of the plant species *Datura innoxia* Mill. obtained when numbers n_{jk} of anthers were prepared under several different conditions. There is one qualitative factor, a treatment consisting of storage at 3 °C for 48 hours or a control storage condition, and a covariate, three values of centrifuging force. We will compare the treatment and control effects on the proportions after adjustment (if necessary) for centrifuging force.

Table 8.5 Anther data

Storage condition		Centrifuging force (g) 40	150	350
Control	y_{1k}	55	52	57
	n_{1k}	102	99	108
Treatment	y_{2k}	55	50	50
	n_{2k}	76	81	90

The proportions $p_{jk} = y_{jk}/n_{jk}$ in the control and treatment groups are plotted against x_k, the logarithm of the centrifuging force, in Fig. 8.4. The response proportions appear to be higher in the treatment group than in the control group and, at least for the treated group, the response decreases with centrifuging force.

We will compare three logistic models for π_{jk}, the probability of the anthers being embriogenic, where $j = 1$ for the control group and $j = 2$ for the treatment group and $x_1 = \log 40 = 3.689$, $x_2 = \log 150 = 5.011$ and $x_3 = \log 350 = 5.858$.

Model 1: $\text{logit}\, \pi_{jk} = \alpha_j + \beta_j x_k$ (i.e. different intercepts and slopes);
Model 2: $\text{logit}\, \pi_{jk} = \alpha_j + \beta x_k$ (i.e. different intercepts but the same slope);
Model 3: $\text{logit}\, \pi_{jk} = \alpha + \beta x_k$ (i.e. same intercept and slope).

These models were fitted by the method of maximum likelihood using GLIM. The results are summarized in Table 8.6.

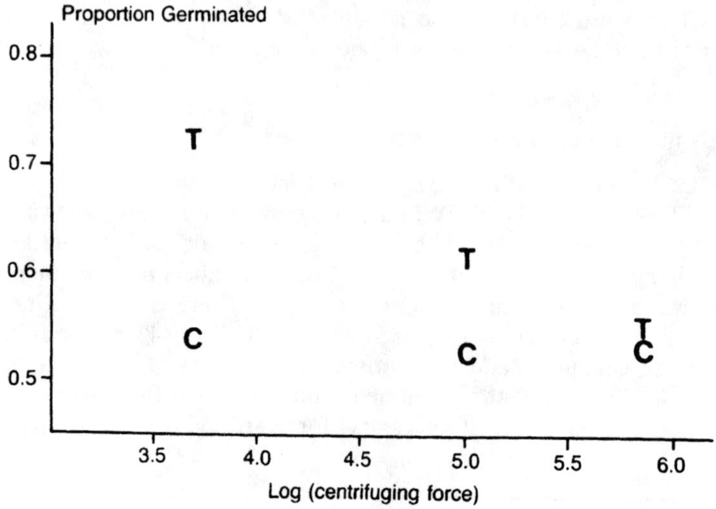

Figure 8.4 Anther data: C and T indicate the control and treatment conditions, respectively.

Table 8.6 Maximum likelihood estimates and log-likelihood ratio statistics for logistic models for the anther data (standard errors of estimates in brackets)

Model 1	Model 2	Model 3
$a_1 = 0.238(0.628)$	$a_1 = 0.877(0.487)$	$a = 1.021(0.481)$
$a_2 - a_1 = 1.977(0.998)$	$a_2 - a_1 = 0.407(0.175)$	$b = -0.148(0.096)$
$b_1 = -0.023(0.127)$	$b = -0.155(0.097)$	
$b_2 - b_1 = -0.319(0.199)$		
$D_1 = 0.0277$	$D_2 = 2.619$	$D_3 = 8.092$

To test the null hypothesis that the slope is the same for the treatment and control groups we use $D_2 - D_1 = 2.591$. From tables for the χ_1^2 distribution, the significance level is between 0.1 and 0.2 and so we could conclude that the data provide little evidence against the null hypothesis of equal slopes. On the other hand, the power of this test is very low and both Fig. 8.4 and the estimates for model 1 suggest that although the slope for the control group may be zero, the slope for the treatment group is negative. Comparison of the log-likelihood statistics from models 2 and 3 gives a test for equality of the control and treatment effects after a common adjustment for centrifuging force: $D_3 - D_2 = 5.473$, which is very significant compared with χ_1^2 distribution so we would conclude that the storage effects differ.

The observed proportions and the corresponding fitted values for models 1 and 2 are shown in Table 8.7. Obviously model 1 fits the data very well but this is hardly surprising since four parameters have been used to describe six data points – such 'over-fitting' is not recommended!

Table 8.7 Observed proportions and probabilities estimated from logistic models for the anther data, and log-likelihood ratio statistics

Storage condition	Covariate value	Observed proportions	Model 1	Model 2
Control	x_1	0.539	0.537	0.576
	x_2	0.525	0.530	0.526
	x_3	0.528	0.525	0.493
Treatment	x_1	0.724	0.721	0.671
	x_2	0.617	0.623	0.625
	x_3	0.555	0.553	0.593
			$D = 0.028$	$D = 2.619$

8.6 OTHER CRITERIA FOR GOODNESS OF FIT

Instead of using maximum likelihood estimation one could estimate the parameters by minimizing the weighted sum of squares

$$S_w = \sum_{i=1}^{N} \frac{(y_i - n_i\pi_i)^2}{n_i\pi_i(1-\pi_i)}$$

since $E(Y_i) = n_i\pi_i$ and $\text{var}(Y_i) = n_i\pi_i(1-\pi_i)$.

This is equivalent to minimizing the **Pearson chi-squared statistic**

$$X^2 = \sum \frac{(o-e)^2}{e}$$

where o represents the observed frequencies in Table 8.1, e represents the expected frequencies obtained from the model and summation is over all $2 \times N$ cells of the table. The reason is

$$X^2 = \sum_{i=1}^{N} \frac{(y_i - n_i\pi_i)^2}{n_i\pi_i} + \sum_{i=1}^{N} \frac{[(n_i - y_i) - n_i(1-\pi_i)]^2}{n_i(1-\pi_i)}$$

$$= \sum_{i=1}^{N} \frac{(y_i - n_i\pi_i)^2}{n_i\pi_i(1-\pi_i)}(1 - \pi_i + \pi_i) = S_w$$

When X^2 is evaluated at the estimated expected frequencies, the statistic is

$$X^2 = \sum_{i=1}^{N} \frac{(y_i - n_i\hat{\pi}_i)^2}{n_i\hat{\pi}_i(1 - \hat{\pi}_i)}$$

which is asymptotically equivalent to the log-likelihood ratio statistics in (8.4).

$$D = 2\sum_{i=1}^{N}\left[y_i \log\left(\frac{y_i}{n_i\hat{\pi}_i}\right) + (n_i - y_i)\log\left(\frac{n_i - y_i}{n_i - n_i\hat{\pi}_i}\right)\right]$$

The proof uses the Taylor series expansion of $s\log(s/t)$ about $s = t$, namely,

$$s\log\frac{s}{t} = (s - t) + \frac{1}{2}\frac{(s - t)^2}{t} + \ldots$$

Thus

$$D = 2\sum_{i=1}^{N}\left\{(y_i - n_i\hat{\pi}_i) + \frac{1}{2}\frac{(y_i - n_i\hat{\pi}_i)^2}{n_i\hat{\pi}_i} + [(n_i - y_i) - (n_i - n_i\hat{\pi}_i)]\right.$$

$$\left. + \frac{1}{2}\frac{[(n_i - y_i) - (n_i - n_i\hat{\pi}_i)]^2}{n_i - n_i\hat{\pi}_i} + \ldots\right\}$$

$$\cong \sum_{i=1}^{N}\frac{(y_i - n_i\hat{\pi}_i)^2}{n_i\hat{\pi}_i(1 - \hat{\pi}_i)} = X^2$$

The large sample distribution of D, under the hypothesis that the model is correct, is $D \sim \chi^2_{N-p}$, therefore approximately $X^2 \sim \chi^2_{N-p}$.

Another criterion for goodness of fit is the **modified chi-squared statistic** obtained by replacing the estimated probabilities in the denominator of X^2 by the relative frequencies,

$$X^2_{\text{mod}} = \sum_{i=1}^{N}\frac{(y_i - n_i\hat{\pi}_i)^2}{y_i(n_i - y_i)/n_i}$$

Asymptotically this too has the χ^2_{N-p} distribution if the model is correct.

The choice between D, X^2 and X^2_{mod} depends on the adequacy of the approximation to the χ^2_{N-p} distribution. There is evidence to suggest that X^2 is often better than D because D is unduly influenced by very small frequencies (Cressie and Read, 1989). All the approximations are likely to be poor if the expected frequencies are too small (e.g. less than 1).

8.7 LEAST SQUARES METHODS

There are some computational advantages in using weighted least squares estimation instead of maximum likelihood, particularly if iteration can be avoided.

Consider a function ψ of the proportion of successes, $P_i = Y_i/n_i$, in the ith subgroup. The Taylor series expansion of $\psi(P_i)$ about $P_i = \pi_i$ is

$$\psi(P_i) = \psi\left(\frac{Y_i}{n_i}\right) = \psi(\pi_i) + \left(\frac{Y_i}{n_i} - \pi_i\right)\psi'(\pi_i) + o\left(\frac{1}{n_i^2}\right)$$

Thus, to a first approximation,

$$E[\psi(P_i)] = \psi(\pi_i)$$

because $E(Y_i/n_i) = \pi_i$. Also

$$\begin{aligned}\text{var}[\psi(P_i)] &= E[\psi(P_i) - \psi(\pi_i)]^2 \\ &= [\psi'(\pi_i)]^2 E\left[\frac{Y_i}{n_i} - \pi_i\right]^2 \\ &= [\psi'(\pi_i)]^2 \frac{\pi_i(1-\pi_i)}{n_i}\end{aligned}$$

because

$$E\left(\frac{Y_i}{n_i} - \pi_i\right)^2 = \text{var}(P_i) = \frac{\pi_i(1-\pi_i)}{n_i}$$

Hence the weighted least squares criterion is

$$X^2 = \sum_{i=1}^{N} \frac{[\psi(y_i/n_i) - \psi(\pi_i)]^2}{[\psi'(\pi_i)]^2 \pi_i(1-\pi_i)/n_i}$$

Some common choices of ψ are summarized in Table 8.8 and discussed below.

Table 8.8 Some weighted least squares models for binary data

$\psi(\pi_i)$	X^2
π_i	$\sum \frac{(p_i - \pi_i)^2}{\pi_i(1-\pi_i)/n_i}$
$\text{logit}\, \pi_i$	$\sum [(\text{logit}\, p_i - \text{logit}\, \pi_i)^2 \pi_i(1-\pi_i)n_i]$
$\arcsin\sqrt{\pi_i}$	$\sum 4n_i[\arcsin\sqrt{p_i} - \arcsin\sqrt{\pi_i}]^2$

First, if $\psi(\pi_i) = \pi_i$ and $\pi_i = \mathbf{x}_i^T\boldsymbol{\beta}$ the modified X^2 criterion is

$$X^2_{\text{mod}} = \sum_{i=1}^{N} \frac{(p_i - \mathbf{x}_i^T\boldsymbol{\beta})^2}{p_i(1-p_i)/n_i} \tag{8.5}$$

which is linear in $\boldsymbol{\beta}$ so estimation does not involve any iteration. However, the estimate $\hat{\pi}_i = \mathbf{x}_i^T\mathbf{b}$ may not lie between 0 and 1.

Second, if $\psi(\pi_i) = \text{logit}\,\pi_i$ and so $\pi_i = \exp(\mathbf{x}_i^{\mathrm{T}}\boldsymbol{\beta})/[1 + \exp(\mathbf{x}_i^{\mathrm{T}}\boldsymbol{\beta})]$ then

$$X^2_{\text{mod}} = \sum_{i=1}^{N} (z_i - \mathbf{x}_i^{\mathrm{T}}\boldsymbol{\beta})^2 \frac{y_i(n_i - y_i)}{n_i} \tag{8.6}$$

where

$$z_i = \text{logit}\,p_i = \log\left(\frac{y_i}{n_i - y_i}\right)$$

This also involves no iteration and yields estimates of the π_i's in the range [0, 1]. Cox (1970) calls this the **empirical logistic transformation** and recommends the use of

$$Z_i = \log\left(\frac{Y_i + \frac{1}{2}}{n_i - Y_i + \frac{1}{2}}\right)$$

instead of

$$Z_i = \log\left(\frac{Y_i}{n_i - Y_i}\right)$$

to reduce the bias $E(Z_i - \mathbf{x}_i^{\mathrm{T}}\boldsymbol{\beta})$ (see Exercise 8.3). The minimum value of (8.6) is called the **minimum logit chi-squared statistic** (Berkson, 1953).

Third, the **arcsin transformation**, $\psi(\pi_i) = \arcsin\sqrt{\pi_i}$ (with any choice of π_i), is said to have the **variance stabilizing property** because

$$\text{var}[\psi(P_i)] = [\psi'(\pi_i)]^2\pi_i(1 - \pi_i)/n_i = (4n_i)^{-1}$$

Thus the weight does not depend on the parameters or the responses and so computations using this transformation are particularly simple and can be performed with a hand calculator.

8.8 REMARKS

Many of the issues that arise in the use of multiple regression for continuous response variables are also relevant with binary responses. Tests for the inclusion or exclusion of certain terms usually are not independent and it is necessary to state carefully which terms are included in the model at each stage. If there are many explanatory variables, stepwise selection methods can be used to identify best subsets of variables.

Graphical examination of residuals is useful for assessing the adequacy of a proposed model. A simple definition for standardized residuals is

$$r_i = \frac{p_i - \hat{\pi}_i}{\sqrt{[(\hat{\pi}_i(1 - \hat{\pi}_i)/n_i]}}$$

where $p_i = y_i/n_i$ is the observed proportion and $\hat{\pi}_i$ is the proportion estimated under the model. The r_i's approximately have a mean of zero and standard deviation of one. They are the signed square roots of contributions to the statistic X^2. When plotted against factor levels and covariates they should not show any systematic patterns. However, their probability distribution may be far from Normal. More complicated residuals, which are nearly Normal, are described by Cox and Snell (1968). More recently Pierce and Schafer (1986) have shown that the signed square roots of contributions to the statistic D,

$$d_i = \pm\sqrt{2}\left[y_i \log\left(\frac{y_i}{n\hat{\pi}_i}\right) + (n_i - y_i)\log\left(\frac{n_i - y_i}{n_i - n_i\hat{\pi}_i}\right)\right]$$

are approximately Normally distributed if the model is good and so they provide appropriate residuals for diagnostic purposes.

8.9 EXERCISES

8.1 Number of deaths from leukaemia and other cancers among survivors of the Hiroshima atom bomb are shown in Table 8.9 classified by the radiation dose received. The data refer to deaths during the period 1950–59 among survivors aged 25–64 years in 1950 (from set 13 of Cox and Snell, 1981, using data from Otake, 1979). Obtain a suitable model to describe the dose–response relationship between radiation and proportional mortality rates for leukaemia.

Table 8.9

	Radiation dose (rads)					
Deaths	0	1–9	10–49	50–99	100–199	200+
Leukaemia	13	5	5	3	4	18
Other cancers	378	200	151	47	31	33
Total cancers	391	205	156	50	35	51

8.2 **Odds ratios**. Consider a 2×2 contingency table from a prospective study in which people who were or were not exposed to some pollutant are followed up and, after several years, categorized according to the presence or absence of a disease. Table 8.10 shows the probabilities for each cell. The odds of disease for either exposure group is $O_i = \pi_i/(1 - \pi_i)$ $(i = 1, 2)$ and so the odds ratio

$$\phi = \frac{O_1}{O_2} = \frac{\pi_1(1 - \pi_2)}{\pi_2(1 - \pi_1)}$$

is a measure of the relative likelihood of disease for the exposed and not exposed groups.

Table 8.10

	Diseased	Not diseased
Exposed	π_1	$1 - \pi_1$
Not exposed	π_2	$1 - \pi_2$

(a) For the simple logistic model $\pi_i = e^{\beta_i}/(1 + e^{\beta_i})$ show that $\phi = 1$ corresponds to no difference between the exposed and not exposed groups.

(b) Consider J 2×2 tables like Table 8.10, one for each level x_j of a factor, such as age group, with $j = 1, \ldots, J$. For the logistic model

$$\pi_{ij} = \frac{\exp(\alpha_i + \beta_i x_j)}{1 + \exp(\alpha_i + \beta_i x_j)} \qquad i = 1, 2; \qquad j = 1, \ldots, J$$

Show that $\log \phi$ is constant over all tables if $\beta_1 = \beta_2$ (McKinlay, 1978).

8.3 Table 8.11 shows numbers of Wisconsin schoolboys planning to attend college cross-classified by:

(a) Socio-economic status (SES) (high H, upper middle UM, lower middle LM, low L);

(b) Intelligence (IQ) (H, UM, LM, L)

(adapted from Example X in Cox and Snell, 1981, using data reported by Sewell and Shah, 1968). Investigate the relationships between socio-economic status and intelligence and plans to attend college first by plotting summary data and then using a suitable program to estimate the magnitudes of effects (the levels of socio-economic status and intelligence can be treated as nominal categories or, using some suitable scoring system, as ordinal categories – see Walter, Feinstein and Wells, 1987).

Table 8.11

SES	IQ	College plans Yes	No	Total	SES	IQ	College plans Yes	No	Total
L	L	17	413	430	UM	L	55	257	312
	LM	42	279	321		LM	80	230	310
	UM	50	180	230		UM	165	192	357
	H	59	110	169		H	204	115	319
LM	L	29	316	345	H	L	43	105	148
	LM	71	296	367		LM	128	137	265
	UM	105	207	312		UM	233	106	339
	H	136	138	274		H	422	71	493

8.4 Let the random variable Y have the binomial distribution with parameters n and π and consider the transformation $\psi[(Y + a)/(n + b)]$ where a and b are constants.

(a) Use the Taylor expansion of $\psi[(Y + a)/(n + b)]$ about $[(Y + a)/(n + b)] = \pi$ and the approximation

$$\frac{Y + a}{n + b} - \pi = \frac{1}{n}[(Y - n\pi) + (a - b\pi)]\left[1 - \frac{b}{n} + \left(\frac{b}{n}\right)^2 - \ldots\right]$$

to show that

$$E\left[\psi\left(\frac{Y + a}{n + b}\right)\right] = \psi(\pi) + \frac{\psi'(\pi)(a - b\pi)}{n} + \frac{\psi''(\pi)\pi(1 - \pi)}{2n} + o\left(\frac{1}{n^2}\right)$$

and

$$\mathrm{var}\left[\psi\left(\frac{Y + a}{n + b}\right)\right] = [\psi'(\pi)]^2\,\frac{\pi(1 - \pi)}{n} + o\left(\frac{1}{n^2}\right)$$

(b) For $\psi(t) = \log[t/(1 - t)]$ show that

$$\text{bias} = E\left[\psi\left(\frac{Y + a}{n + b}\right) - \psi(\pi)\right]$$

is of order n^{-2} if $a = \frac{1}{2}$ and $b = 1$, i.e. the empirical logistic transform $\log[(Y + \frac{1}{2})/(n - Y + \frac{1}{2})]$ is less biased than $\log[Y/(n - Y)]$.

(c) For the log transform $\psi(t) = \log t$ find a and b to reduce the bias to $o(n^{-2})$ and find the variance (Cox and Snell, 1989).

8.5 For the anther data in Table 8.5 fit the linear model

$$\pi_{jk} = \alpha_j + \beta x_k$$

using the modified chi-squared criterion (8.5) and normal regression. Compare the parameter estimates, fitted values and log-likelihood ratio statistic with those obtained using a logistic model $\text{logit}\, \pi_{jk} = \alpha_j + \beta x_k$ (see Tables 8.6 and 8.7).

9

Contingency tables and log-linear models

9.1 INTRODUCTION

This chapter is about the analysis of data in which the response and explanatory variables are all categorical, i.e. they are measured on nominal or possibly ordinal scales. Each scale may have more than two categories. Unlike the methods described in previous chapters, generalized linear models for categorical data can readily be defined when there is more than one variable which can be regarded as a response variable. The observations consist of **counts** or **frequencies** in the cells of a **contingency table** formed by the cross-classification of several variables.

We begin with three numerical examples representing different study designs. For each we consider the roles of the various variables and identify the relevant hypotheses.

Example 9.1 Cross-sectional study of malignant melanoma
These data are from a cross-sectional study of patients with a form of skin cancer called malignant melanoma. For a sample of $n = 400$ patients the site of the tumour and its histological type were recorded. The data, numbers of patients with each combination of tumour type and site, are given in Table 9.1.

Table 9.1 Malignant melanoma: frequencies for tumour type and site (Roberts *et al.*, 1981)

	Site			
Tumour type	Head and neck	Trunk	Extremities	Total
Hutchinson's melanotic freckle	22	2	10	34
Superficial spreading melanoma	16	54	115	185
Nodular	19	33	73	125
Indeterminate	11	17	28	56
Total	68	106	226	400

In this example there are two response variables, site and tumour type. The cell frequencies are regarded as random variables which are subject to the constraint that they must add to n.

The question of interest is whether there is any association between the two response variables. Table 9.2 shows the data displayed as percentages of row and column totals. It appears that Hutchinson's melanotic freckle is more common on the head and neck but there is little evidence of associations between other tumour types and sites.

Table 9.2 Malignant melanoma: row and column percentages for tumour type and site

	Site			
Tumour type	Head and neck	Trunk	Extremities	All sites
Row percentages				
Hutchinson's melanotic freckle	64.7	5.9	29.4	100
Superficial spreading melanoma	8.6	29.2	62.2	100
Nodular	15.2	26.4	58.4	100
Indeterminate	19.6	30.4	50.0	100
All types	17.0	26.5	56.5	100
Column percentages				
Hutchinson's melanotic freckle	32.4	1.9	4.4	8.50
Superficial spreading melanoma	23.5	50.9	50.9	46.25
Nodular	27.9	31.1	32.3	31.25
Indeterminate	16.2	16.0	12.4	14.00
All types	100.0	99.9	100.0	100.0

Example 9.2 Randomized controlled trial of influenza vaccine

In a prospective study of a new living attenuated recombinant vaccine for influenza, patients were randomly allocated to two groups, one of which was given the new vaccine and the other a saline placebo. The responses were titre levels of haemagglutinin inhibiting antibody (HIA) found in the blood six weeks after vaccination.

For this study there is one explanatory variable (the treatment, vaccine or placebo) which is nominal and one response variable (HIA level) which is ordinal but will be treated here as though it too were nominal. The cell frequencies in the rows of Table 9.3 are constrained to add to the numbers of subjects in each treatment group (35 and 38 respectively). We want to know if the pattern of responses is the same for each treatment group.

Table 9.3 Flu vaccine trial (data from R. S. Gillett, personal communication)

	Response			
	Small	Moderate	Large	Total
Placebo	25	8	5	38
Vaccine	6	18	11	35

Example 9.3 Case-control study of gastric and duodenal ulcers and aspirin use

In this retrospective case-control study a group of ulcer patients was assembled and a group of control patients not known to have peptic ulcer who were similar to ulcer patients with respect to age, sex and socio-economic status. Ulcer patients were classified according to the site of the ulcer – gastric or duodenal. Aspirin use was ascertained for all subjects. The results are shown in Table 9.4.

Table 9.4 Gastric and duodenal ulcers and aspirin use: frequencies (Duggan *et al.*, 1986)

	Aspirin use		
	Non-user	User	Total
Gastric ulcer			
Cases	39	25	64
Controls	62	6	68
Duodenal ulcer			
Cases	49	8	57
Controls	53	8	61

This is a $2 \times 2 \times 2$ contingency table with one response variable (aspirin use) and two explanatory variables. In the subtable for each ulcer type the row totals for cases and controls are taken to be fixed. The relevant questions are:

1. Is gastric ulcer associated with aspirin use?
2. Is duodenal ulcer associated with aspirin use?
3. Is any association with aspirin use the same for both ulcer sites?

When the data are presented as percentages of row totals (Table 9.5) it appears that aspirin use is more common among ulcer patients than among controls for gastric ulcer but not for duodenal ulcer.

Table 9.5 Gastric and duodenal ulcers and aspirin use: percentages

	Aspirin use		
	Non-user	User	Total
Gastric ulcer			
Cases	61	39	100
Controls	91	9	100
Duodenal ulcer			
Cases	86	14	100
Controls	87	13	100

This chapter concerns generalized linear models for categorical data when the contingency tables have relatively simple structure. We ignore complicated situations, for example, when some cells of the table necessarily have zero frequencies (e.g. it does not make sense to have any male hysterectomy cases) or when the responses can be regarded as repeated measures on the same individuals. For more complete treatment of contingency tables the reader is referred to the books by Bishop, Fienberg and Holland (1975), Everitt (1977), Fienberg (1980) or Freeman (1987).

9.2 PROBABILITY DISTRIBUTIONS

For two-dimensional tables with J categories for variable A and K categories for variable B we use the notation in Table 9.6 in which Y_{jk} denotes the frequency for the (j, k)th cell, $Y_{j.}$ and $Y_{.k}$ denote the row and column totals and n the overall total. The cell frequencies Y_{jk} are the dependent variables we wish to model.

Table 9.6 Notation for two-dimensional tables

	B_1	B_2	...	B_K	Total
A_1	Y_{11}	Y_{12}	...	Y_{1K}	$Y_{1.}$
A_2	Y_{21}				$Y_{2.}$
$\vdots$	$\vdots$				
A_J	Y_{J1}			Y_{JK}	$Y_{J.}$
Total	$Y_{.1}$	$Y_{.2}$		$Y_{.K}$	$n = Y_{..}$

In general for a $J \times K \times \ldots \times L$ table we write the frequencies $Y_{jk\ldots l}$ in a single vector $\mathbf{y}$ with elements indexed by $i = 1, \ldots, N$.

We begin with probability models for two-dimensional tables. The simplest is obtained by assuming that the random variables Y_{jk} are independent and each has the **Poisson distribution** with parameter $\lambda_{jk} > 0$. Their joint distribution is just the product of the individual Poisson distributions

$$f(\mathbf{y}; \boldsymbol{\lambda}) = \prod_{j=1}^{J} \prod_{k=1}^{K} \frac{\lambda_{jk}^{y_{jk}} e^{-\lambda_{jk}}}{y_{jk}!}$$

More commonly there are constraints on the Y_{jk}'s, for example, that the total frequency n is fixed by the study design. In this case, from the additive property of independent random variables with the Poisson distribution, their sum n also has the Poisson distribution with parameter $\lambda_{..} = \Sigma\Sigma\lambda_{jk}$. Therefore the joint distribution of the Y_{jk}'s, conditional on n, is

$$f(\mathbf{y}|n) = \prod_{j=1}^{J} \prod_{k=1}^{K} \frac{\lambda_{jk}^{y_{jk}} e^{-\lambda_{jk}}}{y_{jk}!} \bigg/ \frac{\lambda_{..}^{n} e^{-\lambda_{..}}}{n!}$$

$$= n! \prod_{j=1}^{J} \prod_{k=1}^{K} \frac{\theta_{jk}^{y_{jk}}}{y_{jk}!} \qquad \text{where } \theta_{jk} = \frac{\lambda_{jk}}{\lambda_{..}}$$

because $\lambda_{..}^{n} = \Pi\Pi\lambda_{jk}^{y_{jk}}$ and $e^{-\lambda_{..}} = \Pi\Pi e^{-\lambda_{jk}}$. This is the **multinomial distribution**. It provides a suitable model for the malignant melanoma data in Example 9.1. By definition $0 \leqslant \theta_{jk} \leqslant 1$ and $\Sigma_j \Sigma_k \theta_{jk} = 1$ and, in fact, the terms θ_{jk} represent the probabilities of the cells.

Another form of constraint applies for tables in which the row or column totals, rather than the overall total, are fixed. In this case the probability distribution for each row (or column) is multinomial; for example, for the jth row with fixed row total $y_{j.}$ the distribution is

$$f(y_{j1}, \ldots, y_{jK}|y_{j.}) = y_{j.}! \prod_{k=1}^{K} \theta_{jk}^{y_{jk}}/y_{jk}!$$

where $\Sigma_k \theta_{jk} = 1$.

The rows (or columns) are assumed to be independent so, for example, if the row totals are fixed, the joint distribution of all the Y_{jk}'s is

$$f(\mathbf{y}|y_{j.}, j = 1, \ldots, J) = \prod_{j=1}^{J} y_{j.}! \prod_{k=1}^{K} \theta_{jk}^{y_{jk}}/y_{jk}!$$

where $\Sigma_k \theta_{jk} = 1$ for each row j. This is the **product multinomial distribution** and it is a suitable model for the randomized controlled trial data in Example 9.2.

For contingency tables with more than two dimensions, if the frequencies are labelled Y_i for $i = 1, \ldots, N$, then the three major probability distributions are as follows.

9.2.1 Poisson distribution

$$f(\mathbf{y}; \boldsymbol{\lambda}) = \prod_{i=1}^{N} \lambda_i^{y_i} e^{-\lambda_i} / y_i! \tag{9.1}$$

with no constraints on the frequencies y_i or on the parameters λ_i.

9.2.2 Multinomial distribution

$$f(\mathbf{y}; \boldsymbol{\theta}|n) = n! \prod_{i=1}^{N} \theta_i^{y_i} / y_i! \tag{9.2}$$

where

$$n = \sum_{i=1}^{N} y_i \quad \text{and} \quad \sum_{i=1}^{N} \theta_i = 1$$

9.2.3 Product multinomial distribution

For a three-dimensional table with J rows, K columns and L layers (subtables), if the row totals are fixed in each layer

$$f(\mathbf{y}; \boldsymbol{\theta}|y_{j.l}, j = 1, \ldots, J; l = 1, \ldots, L) = \prod_{j=1}^{J} \prod_{l=1}^{L} y_{j.l}! \prod_{k=1}^{K} \theta_{jkl}^{y_{jkl}} / y_{jkl}! \tag{9.3}$$

with $\Sigma_k \theta_{jkl} = 1$ for each combination of j and l. If only the layer totals are fixed the distribution is

$$f(\mathbf{y}; \boldsymbol{\theta}|y_{..l}, l = 1, \ldots, L) = \prod_{l=1}^{L} y_{..l}! \prod_{j=1}^{J} \prod_{k=1}^{K} \theta_{jkl}^{y_{jkl}} / y_{jkl}! \tag{9.4}$$

with $\Sigma_j \Sigma_k \theta_{jkl} = 1$ for each l.

The distribution given in (9.3) is a suitable model for the ulcer and aspirin data (Example 9.3) with $J = 2$ for cases or controls, $K = 2$ aspirin use and $L = 2$ for ulcer site.

9.3 LOG-LINEAR MODELS

For the Poisson distribution (9.1) with cell frequencies $Y_1, \ldots, Y_N$ and parameters $\lambda_1, \ldots, \lambda_N$, the expected cell frequencies are given by $E(Y_i) = \lambda_i$.

For any multinomial distribution with cell frequencies $Y_1, \ldots, Y_N$, cell probabilities $\theta_1, \ldots, \theta_N$ with $\Sigma_{i=1}^N \theta_i = 1$ and total frequency $n = \Sigma_{i=1}^N Y_i$, it can be shown that

$$E(Y_i) = n\theta_i \qquad i = 1, \ldots, N \tag{9.5}$$

(e.g. see Bishop, Fienberg and Holland, 1975, Section 13.4).

From result (9.5) it follows that for the product multinomial distributions in (9.3) and (9.4) the expected frequencies are

$$E(Y_{jkl}) = y_{j.l}\theta_{jkl} \tag{9.6}$$

and

$$E(Y_{jkl}) = y_{..l}\theta_{jkl}$$

respectively.

For two-dimensional contingency tables (Table 9.6) all the usual hypotheses can be formulated as multiplicative models for the expected cell frequencies. For example, if the hypothesis is that the row and column variables are independent then $\theta_{jk} = \theta_{j.}\theta_{.k}$ where $\theta_{j.}$ and $\theta_{.k}$ represent the marginal probabilities of the row and column variables and $\Sigma_j \theta_{j.} = 1$ and $\Sigma_k \theta_{.k} = 1$. Hence for the multinomial distribution, from (9.5), the expected frequencies are

$$E(Y_{jk}) = n\theta_{j.}\theta_{.k} \tag{9.7}$$

For a two-dimensional table with fixed row totals $y_{j.}$, the hypothesis that the cell probabilities are the same in all rows, called the **homogeneity hypothesis**, can be written as $\theta_{jk} = \theta_{.k}$ for all j. Therefore for the product multinomial distribution the expected frequencies are

$$E(Y_{jk}) = y_{j.}\theta_{.k}$$

with $\Sigma_k \theta_{.k} = 1$.

Similarly, for tables in higher dimensions the most common hypotheses can be expressed as multiplicative models in which the expected cell frequencies are given by products of marginal probabilities and fixed marginal total frequencies.

This suggests that for generalized linear models the logarithm is the natural link function between $E(Y_i)$ and a linear combination of parameters, i.e.

$$\eta_i = \log E(Y_i) = \mathbf{x}_i^{\mathrm{T}}\boldsymbol{\beta} \qquad i = 1, \ldots, N$$

hence the name **log-linear model**. For example, (9.7) can be expressed as

$$\eta_{jk} = \log E(Y_{jk}) = \mu + \alpha_j + \beta_k \tag{9.8}$$

and, by analogy with analysis of variance, the corresponding maximal model $E(Y_{jk}) = n\theta_{jk}$ can be written as

$$\eta_{jk} = \log E(Y_{jk}) = \mu + \alpha_j + \beta_k + (\alpha\beta)_{jk} \qquad (9.9)$$

so that the independence hypothesis $\theta_{jk} = \theta_{j.}\theta_{.k}$ for all j and k is equivalent to the 'no interaction' hypothesis that $(\alpha\beta)_{jk} = 0$ for all j and k.

The higher-order terms of log-linear models are usually defined as deviations from lower-order terms. For example, in (9.8) α_j represents the differential effect of row j beyond the average effect μ. Also the models are **hierarchical** in the sense that higher-order terms are not included in a model unless all the related lower-order terms are included.

As with ANOVA models the log-linear models (9.8) and (9.9) have too many parameters so that sum-to-zero or corner-point constraints are needed. In general, for **main effects** α_j where $j = 1, \ldots, J$ there are $(J - 1)$ independent parameters; for **first-order interactions** $(\alpha\beta)_{jk}$ where $j = 1, \ldots, J$ and $k = 1, \ldots, K$, there are $(J - 1)(K - 1)$ independent parameters, and so on.

In the analysis of contingency table data the main questions almost always relate to associations between variables. Therefore in log-linear models the terms of primary interest are the interactions involving two or more variables. As the models are hierarchical this means that models used for hypothesis testing involve interaction terms and all the corresponding main effects.

In the expressions for expected cell frequencies for multinomial and product multinomial distributions certain terms are fixed constants, for instance n in (9.5) or $y_{j.l}$ in (9.6). This means that the corresponding parameters must always be included in the log-linear models. For example, the maximal model corresponding to $E(Y_{jkl}) = y_{j.l}\theta_{jkl}$ in (9.6) is

$$\eta_{jkl} = \mu + \alpha_j + \beta_k + \gamma_l + (\alpha\beta)_{jk} + (\alpha\gamma)_{jl} + (\beta\gamma)_{kl} + (\alpha\beta\gamma)_{jkl}$$

in which the expression

$$\mu + \alpha_j + \gamma_l + (\alpha\gamma)_{jl} \qquad (9.10)$$

corresponds to the fixed marginal total $y_{j.l}$ and the remainder

$$\beta_k + (\alpha\beta)_{jk} + (\beta\gamma)_{kl} + (\alpha\beta\gamma)_{jkl} \qquad (9.11)$$

corresponds to the cell probability θ_{jkl}. Thus any hypothesis about the structure of the cell probabilities is formulated by omitting terms from (9.11) while the expression (9.10) is a necessary part of any model.

Table 9.7 summarizes the most commonly used log-linear models for

Table 9.7 Log-linear models for two-dimensional contingency tables

Log-linear model	Poisson distribution	Multinominal distribution	Product multinomial distribution with $y_{j.}$ fixed
Maximal model $\mu + \alpha_j + \beta_k + (\alpha\beta)_{jk}$ with JK independent parameters	$E(Y_{jk}) = \lambda_{jk}$	$E(Y_{jk}) = n\theta_{jk}$ with $\Sigma_j \Sigma_k \theta_{jk} = 1$	$E(Y_{jk}) = y_{j.}\theta_{jk}$ with $\Sigma_k \theta_{jk} = 1$ for $j = 1, \ldots, J$.
$\mu + \alpha_j + \beta_k$ with $J + K - 1$ independent parameters	Independence hypothesis $E(Y_{jk}) = \lambda_j \lambda_k$	Independence hypothesis $E(Y_{jk}) = n\theta_{j.}\theta_{.k}$ with $\Sigma_j \theta_{j.} = \Sigma_k \theta_{.k} = 1$	Homogeneity hypothesis $E(Y_{jk}) = y_{j.}\theta_{.k}$ with $\Sigma_k \theta_{.k} = 1$
Terms which must be included in any log-linear model		μ since n is fixed	$\mu + \alpha_j$ since $y_{j.}$ is fixed

two-dimensional contingency tables. Generally the same models apply for all three probability distributions although there are differences in the terms which must be included in the models and differences in the interpretation of the 'interaction' term. For three-dimensional tables, models corresponding to the major hypotheses for multinomial and product multinomial distributions are given in Appendix D.

9.4 MAXIMUM LIKELIHOOD ESTIMATION

For the Poisson distribution (9.1) the log-likelihood function is

$$l = \sum_i (y_i \log \lambda_i - \lambda_i - \log y_i!)$$

where $\lambda_i = E(Y_i)$. For the multinomial distribution (9.2) the log-likelihood function is

$$l = \log n! + \sum_i (y_i \log \theta_i - \log y_i!)$$

which can be written in the form

$$l = \text{constant} + \sum_i y_i \log E(Y_i)$$

because, by (9.5), $E(Y_i) = n\theta_i$ (subject to the constraint $\Sigma \theta_i = 1$). Similarly for any product multinomial distribution the log-likelihood function can be written as

$$l = \text{constant} + \sum_i y_i \log E(Y_i)$$

where the $E(Y_i)$'s are subject to various constraints. Thus for all three probability distributions the log-likelihood function depends only on the observed cell frequencies $\mathbf{y}$ and their expected frequencies $E(\mathbf{y})$.

There are two approaches to estimating the expected frequencies. One is to estimate them directly by maximizing the log-likelihood function subject to the relevant constraints. The other, consistent with the usual approach to generalized linear models, is to use the log-linear model

$$\eta_i = \log E(Y_i) = \boldsymbol{x}_i^T \boldsymbol{\beta}$$

first estimating $\boldsymbol{\beta}$ and then using the estimates to calculate the η_is and hence the fitted values $\exp(\eta_i)$. By the invariance property of maximum likelihood estimators these fitted values will be the maximum likelihood estimates of the expected frequencies $E(Y_i)$.

Birch (1963) showed that for any log-linear model the maximum likelihood estimators are the same for all three probability distributions, provided that the parameters which correspond to the fixed marginal totals are always included in the model (as discussed in section 9.3). This means that for the purpose of estimation the Poisson distribution can be assumed. As the Poisson distribution belongs to the exponential family and constraints on parameters in the log-linear models $\boldsymbol{\eta} = \boldsymbol{X\beta}$ can be accommodated by suitable choice of the elements of $\boldsymbol{\beta}$, all the standard results for generalized linear models apply. In particular, the Newton–Raphson estimation procedures described in Chapter 5 may be used to estimate $\boldsymbol{\beta}$. This approach has been advocated by Nelder (1974) and is implemented in GLIM.

The alternative approach, based on estimating the expected cell frequencies $E(Y_i)$ directly, is to obtain the maximum likelihood estimators in such a way as to incorporate any constraints on the probabilities. For the expected cell frequencies explicit closed-form solutions of the restricted maximum likelihood equations may not exist so approximate numerical solutions have to be calculated. An iterative method is used to adjust the estimated expected cell frequencies until they add up to the required marginal totals (at least to within some specified accuracy). This procedure is called **iterative proportional fitting**. It is described in detail by Bishop, Fienberg and Holland (1975) and it is implemented in many statistical computing programs (for example, P4F in BMDP).

Maximum likelihood estimation is illustrated by numerical examples in section 9.6 after hypothesis testing has been considered.

9.5 HYPOTHESIS TESTING AND GOODNESS OF FIT

For the maximal model in which there are N parameters these can be taken, without loss of generality, to be the expected frequencies $E(Y_1), \ldots, E(Y_N)$. They can be estimated by the corresponding observed frequencies $y_1, \ldots, y_N$. So for any of the three probability distributions, the log-likelihood function has the form

$$l(\mathbf{b}_{\max}; \mathbf{y}) = \text{constant} + \sum_i y_i \log y_i$$

For any other model let e_i denote the estimated expected cell frequencies so that the log-likelihood function is

$$l(\mathbf{b}; \mathbf{y}) = \text{constant} + \sum_i y_i \log e_i$$

where the constants are the same. Hence the log-likelihood ratio statistic is

$$D = 2[l(\mathbf{b}_{\max}; \mathbf{y}) - l(\mathbf{b}; \mathbf{y})] = 2 \sum_{i=1}^{N} y_i \log \frac{y_i}{e_i}$$

which is of the form

$$D = 2 \sum o \log \frac{o}{e}$$

where o and e denote the observed and estimated expected (i.e. fitted) cell frequencies respectively, and summation is over all cells in the table. From Chapter 5, if the model fits the data well, then for large samples D has the central chi-squared distribution with degrees of freedom given by the **number of cells with non-zero observed frequencies** (i.e. N if $y_i > 0$ for all i) **minus the number of independent, non-zero parameters** in the model.

The chi-squared statistic

$$X^2 = \sum \frac{(o - e)^2}{e}$$

is more commonly used for contingency table data than D. By the argument used in section 8.6, it can readily be shown that these two statistics are asymptotically equivalent and hence that, for large samples, X^2 has the chi-squared distribution with the number of degrees of freedom given above.

The form of the chi-squared statistic suggests that the standardized residual for each cell can be defined as

$$r_i = \frac{o_i - e_i}{\sqrt{e_i}}$$

This definition also follows naturally from the Poisson model because $E(Y_i) = \text{var}(Y_i)$ and so e_i is also an estimate of the variance of the cell frequency. Departures from the model may be detected by inspecting the residuals. Values which are too far from zero in either direction (say $|r_i| > 3$ corresponding roughly to the 1% tails of the standard Normal distribution) or patterns in the residuals from certain parts of the table may suggest other, more appropriate models.

9.6 NUMERICAL EXAMPLES

9.6.1 Cross-sectional study of malignant melanoma (Example 9.1)

We want to investigate whether there is any association between tumour type and site. We do this by testing the null hypothesis that the two variables are independent

$$H_0 : E(Y_{jk}) = n\theta_{j.}\theta_{.k}$$

where $\Sigma\,\theta_{j.} = 1$ and $\Sigma\,\theta_{.k} = 1$. The corresponding log-linear model is

$$\eta_{jk} = \log E(Y_{jk}) = \mu + \alpha_j + \beta_k$$

subject to constraints such as

$$\sum \alpha_j = 0 \quad \text{and} \quad \sum \beta_k = 0 \text{ (sum to zero constraints)}$$

or

$$\alpha_1 = 0 \quad \text{and} \quad \beta_1 = 0 \text{ (corner-point constraints)}$$

As there are $J = 4$ tumour types and $K = 3$ sites there are $1 + (J - 1) + (K - 1) = J + K - 1 = 6$ parameters to be estimated. This model is (implicitly) compared with the maximal model

$$\eta_{jk} = \mu + \alpha_j + \beta_k + (\alpha\beta)_{jk}$$

(subject to appropriate constraints). The maximal model has $p = 12$ parameters so that $\hat{\eta}_{jk} = \log y_{jk}$ and $D = 0$. If H_0 is correct then the test statistics D or X^2 have the distribution χ^2_{N-p} where $N = 12$ and $p = 6$.

Table 9.8 shows the analysis of these data using GLIM. The Poisson distribution is used with the logarithmic link function. The parameters correspond to the corner-point constraints. The fitted values are obtained from the estimates as shown by the following examples:

$$\hat{\eta}_{11} = 1.754 \quad \text{so} \quad e_{11} = e^{1.754} = 5.78$$

$$\hat{\eta}_{43} = 1.754 + 0.499 + 1.201 = 3.454 \quad \text{so} \quad e_{43} = e^{3.454} = 31.64$$

The log-likelihood ratio statistic is $D = 2\Sigma\, o \log(o/e) = 51.795$. The

Table 9.8 Analysis of malignant melanoma data using GLIM (version 3.77)

```
? $units 12 $
? $factors type 4 site 3 $
? $data y $
? $read
$REA? 22  2  10
$REA? 16 54 115
$REA? 19 33  73
$REA? 11 17  28
? $calc type = %gl(4, 3) $
? $calc site  = %gl(3, 1) $
? $yvar y $
? $error poisson $
? $link log $
? $fit type + site $
 scaled deviance= 51.795 at cycle 4
            d.f.= 6
? $dis e r$
          estimate        s.e.        parameter
  1       1.754           0.2036      1
  2       1.694           0.1862      TYPE(2)
  3       1.302           0.1930      TYPE(3)
  4       0.4990          0.2170      TYPE(4)
  5       0.4439          0.1553      SITE(2)
  6       1.201           0.1383      SITE(3)
  scale parameter taken as 1.000
unit      observed        fitted       residual
  1          22            5.780        6.747
  2           2            9.010       -2.335
  3          10           19.210       -2.101
  4          16           31.450       -2.755
  5          54           49.025        0.711
  6         115          104.525        1.025
  7          19           21.250       -0.488
  8          33           33.125       -0.022
  9          73           70.625        0.283
 10          11            9.520        0.480
 11          17           14.840        0.561
 12          28           31.640       -0.647
```

chi-squared statistic is $X^2 = \Sigma[(o - e)^2/e] = 65.813$. From either statistic it is apparent that the model fits poorly since $\Pr(\chi_6^2 > 50) < 0.001$. So we reject the independence hypothesis H_0 and conclude that there is some association between tumour type and site. The residuals are given by $r = (o - e)/\sqrt{e}$. The largest residual, observation 1 for the cell (1, 1), accounts for much of the lack of fit, confirming that the main 'signal' in the data is the association of Hutchinson's melanotic freckle with the head and neck.

For the alternative estimation strategy we first obtain the fitted values e_{jk} and then use them to estimate the parameters of the log-linear model. The log-likelihood function based on the multinomial distribution is

$$l = \log n! + \sum \sum (y_{jk} \log \theta_{jk} - \log y_{jk}!)$$

If H_0 is true then $\theta_{jk} = \theta_{j.}\theta_{.k}$. One way to maximize the resulting log-likelihood function subject to the constraints $\Sigma\, \theta_{j.} = 1$ and $\Sigma\, \theta_{.k} = 1$ is to use Lagrange multipliers u and v and maximize the function

$$t = \text{constant} + \sum \sum [y_{jk} \log (\theta_{j.}\theta_{.k})] - u\left(\sum \theta_{j.} - 1\right) - v\left(\sum \theta_{.k} - 1\right)$$

with respect to $\theta_{j.}$ ($j = 1, \ldots, J$), $\theta_{.k}$ ($k = 1, \ldots, K$), u and v.

The solutions of the equations

$$\frac{\partial t}{\partial \theta_{j.}} = \sum_k \frac{y_{jk}}{\theta_{j.}} - u = 0 \qquad \frac{\partial t}{\partial \theta_{.k}} = \sum_j \frac{y_{jk}}{\theta_{.k}} - v = 0$$

$$\frac{\partial t}{\partial u} = -\sum_j \theta_{j.} + 1 = 0 \qquad \frac{\partial t}{\partial v} = -\sum_k \theta_{.k} + 1 = 0$$

are obtained from

$$\sum_k y_{jk} = \hat{u}\hat{\theta}_{j.} \qquad \sum_j y_{jk} = \hat{v}\hat{\theta}_{.k} \qquad \sum_j \hat{\theta}_{j.} = 1 \quad \text{and} \quad \sum_k \hat{\theta}_{.k} = 1$$

so that

$$\hat{u} = \sum_j \sum_k y_{jk} = n \qquad \hat{v} = \sum_j \sum_k y_{jk} = n \qquad \hat{\theta}_{j.} = \sum_k y_{jk}/n = y_{j.}/n$$

and

$$\hat{\theta}_{.k} = \sum_j y_{jk}/n = y_{.k}/n$$

Therefore

$$e_{jk} = y_{j.}y_{.k}/n$$

Now the equations

$$\hat{\eta}_{jk} = \log e_{jk} = \hat{\mu} + \hat{\alpha}_j + \hat{\beta}_k$$

with $\Sigma\, \hat{\alpha}_j = 0$ and $\Sigma\, \hat{\beta}_k = 0$ can be used to obtain the estimates

$$\hat{\mu} = \frac{1}{N}\sum\sum \hat{\eta}_{jk} = \frac{1}{J}\sum \log y_{j.} + \frac{1}{K}\sum \log y_{.k} - \log n$$

$$\hat{\alpha}_j = \log y_{j.} - \frac{1}{J}\sum \log y_{j.}$$

and

$$\hat{\beta}_k = \log y_{.k} - \frac{1}{K}\sum \log y_{.k}$$

For the malignant melanoma data the estimates are

$$\hat{\mu} = 3.176 \quad \hat{\alpha}_1 = -0.874 \quad \hat{\alpha}_2 = 0.820 \quad \hat{\alpha}_3 = 0.428$$

$$\hat{\alpha}_4 = -0.375 \quad \hat{\beta}_1 = -0.548 \quad \hat{\beta}_2 = -0.104 \text{ and } \hat{\beta}_3 = 0.653$$

These give the same values for the fitted values as the model in Table 9.8; for example for the first and last observations

$$\hat{\eta}_{11} = \hat{\mu} + \hat{\alpha}_1 + \hat{\beta}_1 = 1.754 \text{ so } e_{11} = e^{1.754} = 5.78$$

$$\hat{\eta}_{43} = \hat{\mu} + \hat{\alpha}_4 + \hat{\beta}_3 = 3.454 \text{ so } e_{43} = e^{3.454} = 31.64$$

9.6.2 Gastric and duodenal ulcers and aspirin use (Example 9.3)

To investigate whether gastric ulcer is associated with aspirin use we test the null hypothesis that the probability of aspirin use is independent of disease status (case or control) for the gastric ulcer group ($l = 1$) against the alternative hypothesis of non-independence:

$$H_0 : E(Y_{jkl}) = y_{j.l}\theta_{.kl}$$

$$H_1 : E(Y_{jkl}) = y_{j.l}\theta_{jkl}$$

with the row totals $y_{j.l}$ taken as fixed, see (9.3). Analogous hypotheses can be specified for duodenal ulcer (using $l = 2$). So the joint hypothesis of no association between aspirin and disease, for either ulcer site, corresponds to the log-linear model

$$\eta_{jkl} = \log E(Y_{jkl}) = \mu + \alpha_j + \gamma_l + (\alpha\gamma)_{jl} + \beta_k + (\beta\gamma)_{kl} \quad (9.12)$$

where the first four terms of (9.12) correspond to the fixed row totals $y_{j.l}$ and the last two terms cover the aspirin effect, allowing for possibly different levels of aspirin use for each ulcer site. This model is compared with the maximal model to test the hypothesis of no association between aspirin use and ulcer. If this hypothesis is rejected we can test the hypothesis that the extent of association is the same for both ulcer sites using a multiplicative probability term $\theta_{jk.}$ or, equivalently, the log-linear model

$$\eta_{jkl} = \mu + \alpha_j + \gamma_l + (\alpha\gamma)_{jl} + \beta_k + (\beta\gamma)_{kl} + (\alpha\beta)_{jk} \qquad (9.13)$$

Table 9.9 shows the results of fitting models (9.12) and (9.13) using GLIM. For model (9.12) the log-likelihood ratio statistic, $D = 17.697$ with 2 degrees of freedom, indicates a poor fit so we would reject the hypothesis of no association between aspirin use and gastric or duodenal ulcers. Model (9.13) is significantly better, $\Delta D = 11.41$ with 1 degree of freedom, confirming the existence of an association. Nevertheless it is still poor, $D = 6.283$ with 1 degree of freedom, so we would reject the hypothesis that the association with aspirin use is the same for gastric ulcer and duodenal ulcer. This is consistent with the remark about Table 9.5 that aspirin use seems to be associated with gastric ulcer but not duodenal ulcer.

Table 9.9 Analysis of the ulcer and aspirin data using GLIM (version 3.77)

```
? $units 8 $
? $factors cascon 2 aspirin 2 site 2 $
? $data y cascon aspirin site $
? $read
$REA? 39 1 1 1
$REA? 25 1 2 1
$REA? 62 2 1 1
$REA?  6 2 2 1
$REA? 49 1 1 2
$REA?  8 1 2 2
$REA? 53 2 1 2
$REA?  8 2 2 2
? $yvar y $
? $error poisson $
? $link log $
? $fit cascon + site + cascon.site + aspirin + aspirin.site $
 scaled deviance = 17.697 at cycle 4
            d.f. = 2
? $fit + aspirin.cascon $
 scaled deviance = 6.2830  (change = −11.41) at cycle 4
            d.f. = 1       (change = −1     )
```

9.7 REMARKS

The numerical examples in section 9.6 are particularly simple and the calculations and the interpretation of the results are straightforward. For contingency tables involving more than three variables, model selection and interpretation become much more complicated. Some suggestions for systematically fitting complex log-linear models are given by Bishop,

Fienberg and Holland (1975, Chs 4 and 9), and Whittaker and Aitkin (1978). The analysis of multidimensional contingency tables usually requires a computer to perform the iterative estimation.

An alternative approach to the likelihood methods considered in this chapter has been proposed by Grizzle, Starmer and Koch (1969). It is based on modelling functions of the multinomial probabilities $\boldsymbol{\theta}$ as linear combinations of parameters, i.e.

$$\boldsymbol{F}(\boldsymbol{\theta}) = \boldsymbol{X}\boldsymbol{\beta}$$

and using the weighted least squares criterion

$$S_w = [\boldsymbol{F}(\mathbf{p}) - \boldsymbol{X}\boldsymbol{\beta}]^T \boldsymbol{V}^{-1}[\boldsymbol{F}(\mathbf{p}) - \boldsymbol{X}\boldsymbol{\beta}]$$

for estimation and hypothesis testing (where $\mathbf{p}$ represents the estimated probabilities and $\boldsymbol{V}$ the variance–covariance matrix for $\boldsymbol{F}(\mathbf{p})$). An advantage of this method is that it can be used for linear and non-linear (including log-linear) models. But it is computationally more complex than the likelihood methods and is less widely used. For an introduction to this approach, see Freeman (1987).

Contingency table methods, including log-linear models, are primarily designed for analysing data for nominal categories. In practice they are also used for ordinal categories, either ignoring the ordering or assigning covariate scores to the categories, see Walter, Feinstein and Wells (1987). McCullagh (1980) has shown that generalized linear modelling can be extended to give regression-like methods for ordinal data. The details are beyond the scope of this book and the reader is referred to the original paper or McCullagh and Nelder (1989).

9.8 EXERCISES

9.1 For the randomized controlled trial of influenza vaccine, Example 9.2:

(a) Test the hypothesis that the response pattern is the same for the placebo and vaccine groups;
  (i) using the usual methods for $r \times s$ contingency tables
  (ii) using log-linear models.

(b) For the model corresponding to no differences in response:
  (i) calculate the standardized residuals – do they show any patterns which are useful for interpreting the hypothesis test in (a)?
  (ii) calculate and compare the test statistics D and X^2.

9.2 The data in Table 9.10 relate to an investigation into satisfaction with housing conditions in Copenhagen (derived from Example W of Cox and Snell, 1981, from original data from Madsen, 1976). Residents of selected areas living in rented homes built between 1960 and 1968 were questioned about their satisfaction and the degree of contact with other residents. The data were tabulated by type of housing. Investigate the associations between satisfaction, contact with other residents and type of housing.

(a) Produce appropriate tables of percentages to gain initial insights into the data; for example, percentages in each contact category by type of housing, or percentages in each category of satisfaction by contact and type of housing.
(b) Use an appropriate statistical computing program to fit log-linear models to investigate interactions among the variables.
(c) For some model that fits (at least moderately) well, calculate the standardized residuals and use them to find where largest discrepancies are between the observed and expected values.

Table 9.10

	Contact with other residents					
	Low			High		
Satisfaction	Low	Medium	High	Low	Medium	High
Tower blocks	65	54	100	34	47	100
Apartments	130	76	111	141	116	191
Houses	67	48	62	130	105	104

9.3 For a 2 × 2 contingency table the maximal log-linear model can be written as

$$\eta_{11} = \mu + \alpha + \beta + (\alpha\beta) \qquad \eta_{12} = \mu + \alpha - \beta - (\alpha\beta)$$

$$\eta_{21} = \mu - \alpha + \beta - (\alpha\beta) \qquad \eta_{22} = \mu - \alpha - \beta + (\alpha\beta)$$

where $\eta_{jk} = \log E(Y_{jk}) = \log(n\theta_{jk})$ and $n = \Sigma\Sigma\, Y_{jk}$.
Show that the 'interaction' term $(\alpha\beta)$ is given by

$$(\alpha\beta) = \tfrac{1}{4}\log\phi$$

where ϕ is the odds ratio $(\theta_{11}\theta_{22})/(\theta_{12}\theta_{21})$ and hence that $\phi = 1$ corresponds to no interaction.

9.4 Consider a $2 \times K$ contingency table (Table 9.11) in which the column totals $y_{.k}$ are fixed for $k = 1, \ldots, K$.

(a) Show that the product multinomial distribution for this table reduces to

$$f(z_1, \ldots, z_K | n_1, \ldots, n_K) = \prod_{k=1}^{K} \binom{n_k}{z_k} \pi_k^{z_k}(1 - \pi_k)^{n_k - z_k}$$

where $n_k = y_{.k}$, $z_k = y_{1k}$, $n_k - z_k = y_{2k}$, $\pi_k = \theta_{1k}$ and $1 - \pi_k = \theta_{2k}$ (for $k = 1, \ldots, K$). This is the **product binomial distribution** and is the joint distribution for Table 8.1 (with appropriate changes in notation).

Table 9.11

	1	...	k	...	K
Success	y_{11}	...	y_{1k}	...	y_{1K}
Failure	y_{21}	...	y_{2k}	...	y_{2K}
Total	$y_{.1}$	...	$y_{.k}$	...	$y_{.K}$

(b) Show that the log-linear model with

$$\eta_{1k} = \log E(Z_k) = \mathbf{x}_{1k}^{\mathrm{T}}\boldsymbol{\beta}$$

and

$$\eta_{2k} = \log E(n_k - Z_k) = \mathbf{x}_{2k}^{\mathrm{T}}\boldsymbol{\beta}$$

is equivalent to the logistic model

$$\log\left(\frac{\pi_k}{1 - \pi_k}\right) = \mathbf{x}_k^{\mathrm{T}}\boldsymbol{\beta}$$

where $\mathbf{x}_k = \mathbf{x}_{1k} - \mathbf{x}_{2k}$.

(c) Analyse the data on aspirin use and gastric and duodenal ulcers (Example 9.3) using logistic regression and compare the results from those obtained in section 9.6.2.

Appendix A

Consider a continuous random variable Y with probability density function $f(y;\,\theta)$ depending on a single parameter θ (or if Y is discrete $f(y;\,\theta)$ is its probability distribution). The log-likelihood function is the logarithm of $f(y;\,\theta)$ regarded primarily as a function of θ, i.e.

$$l(\theta;\,y) = \log f(y;\,\theta)$$

Many of the key results about generalized linear models relate to the derivative

$$U = \frac{\mathrm{d}l}{\mathrm{d}\theta}$$

which is called the **score**.

To find the moments of U we use the identity

$$\frac{\mathrm{d}\log f(y;\,\theta)}{\mathrm{d}\theta} = \frac{1}{f(y;\,\theta)}\,\frac{\mathrm{d}f(y;\,\theta)}{\mathrm{d}\theta} \tag{A.1}$$

If we take expectations of (A.1) we obtain

$$E(U) = \int \frac{\mathrm{d}\log f(y;\,\theta)}{\mathrm{d}\theta} f(y;\,\theta)\,\mathrm{d}y = \int \frac{\mathrm{d}f(y;\,\theta)}{\mathrm{d}\theta}\,\mathrm{d}y$$

where integration is over the domain of y. Under certain regularity conditions the right-hand term is

$$\int \frac{\mathrm{d}f(y;\,\theta)}{\mathrm{d}\theta}\,\mathrm{d}y = \frac{\mathrm{d}}{\mathrm{d}\theta}\int f(y;\,\theta)\,\mathrm{d}y = \frac{\mathrm{d}}{\mathrm{d}\theta}1 = 0$$

since $\int f(y;\,\theta)\,\mathrm{d}y = 1$. Hence

$$E(U) = 0 \tag{A.2}$$

Also if we differentiate (A.1) with respect to θ and take expectations, provided the order of these operations can be interchanged, then

$$\frac{\mathrm{d}}{\mathrm{d}\theta}\int \frac{\mathrm{d}\log f(y;\,\theta)}{\mathrm{d}\theta} f(y;\,\theta)\,\mathrm{d}y = \frac{\mathrm{d}^2}{\mathrm{d}\theta^2}\int f(y;\,\theta)\,\mathrm{d}y$$

The right-hand side equals zero because $\int f(y;\,\theta)\,\mathrm{d}y = 1$ and the left-hand side can be expressed as

$$\int \frac{d^2 \log f(y;\theta)}{d\theta^2} f(y;\theta)\,dy + \int \frac{d \log f(y;\theta)}{d\theta} \frac{df(y;\theta)}{d\theta}\,dy$$

Hence, substituting (A.1) in the second term, we obtain

$$\int \frac{d^2 \log f(y;\theta)}{d\theta^2} f(y;\theta)\,dy + \int \left[\frac{d \log f(y;\theta)}{d\theta}\right]^2 f(y;\theta)\,dy = 0$$

Therefore

$$E\left[-\frac{d^2 \log f(y;\theta)}{d\theta^2}\right] = E\left\{\left[\frac{d \log f(y;\theta)}{d\theta}\right]^2\right\}$$

In terms of the score statistic this is

$$E(-U') = E(U^2)$$

where U' denotes the derivative of U with respect to θ. Since $E(U) = 0$, the variance of U, which is called the **information**, is

$$\operatorname{var}(U) = E(U^2) = E(-U') \qquad \text{(A.3)}$$

More generally consider independent random variables $Y_1, \ldots, Y_N$ whose probability distributions depend on parameters $\theta_1, \ldots, \theta_p$ where $p \leqslant N$. Let $l_i(\boldsymbol{\theta}; y_i)$ denote the log-likelihood function of Y_i where $\boldsymbol{\theta}$ is the vector of $\theta_1, \ldots, \theta_p$. Then the log-likelihood function of $Y_1, \ldots, Y_N$ is

$$l(\boldsymbol{\theta}; \mathbf{y}) = \sum_{i=1}^{N} l_i(\boldsymbol{\theta}; y_i)$$

where $\mathbf{y} = [Y_1, \ldots, Y_N]^T$. The **total score** with respect to θ_j is defined as

$$U_j = \frac{\partial l(\boldsymbol{\theta}; \mathbf{y})}{\partial \theta_j} = \sum_{i=1}^{N} \frac{\partial l_i(\boldsymbol{\theta}; y_i)}{\partial \theta_j}$$

By the same argument as for (A.2),

$$E\left[\frac{\partial l_i(\boldsymbol{\theta}; y_i)}{\partial \theta_j}\right] = 0$$

and so

$$E(U_j) = 0 \quad \text{for all } j \qquad \text{(A.4)}$$

The **information matrix** is defined to be the variance–covariance matrix of the U_js, $\mathcal{J} = E(\mathbf{U}\mathbf{U}^T)$ where $\mathbf{U}$ is the vector of $U_1, \ldots, U_p$, so it has elements

$$\mathcal{J}_{jk} = E[U_j U_k] = E\left[\frac{\partial l}{\partial \theta_j} \frac{\partial l}{\partial \theta_k}\right] \qquad \text{(A.5)}$$

By an argument analogous to that for the single random variable and single parameter case above, it can be shown that

$$E\left[\frac{\partial l_i}{\partial \theta_j}\frac{\partial l_i}{\partial \theta_k}\right] = E\left[-\frac{\partial^2 l_i}{\partial \theta_j \partial \theta_k}\right]$$

Hence the elements of the information matrix are also given by

$$\mathcal{I}_{jk} = E\left[-\frac{\partial^2 l}{\partial \theta_j \partial \theta_k}\right] \tag{A.6}$$

Appendix B

From sections 3.2 and 3.3, for the generalized linear model the log-likelihood function can be written as

$$l(\boldsymbol{\theta}; \mathbf{y}) = \sum y_i b(\theta_i) + \sum c(\theta_i) + \sum d(y_i)$$

with

$$E(Y_i) = \mu_i = -c'(\theta_i)/b'(\theta_i) \tag{A.7}$$

and

$$g(\mu_i) = \mathbf{x}_i^{\mathrm{T}}\boldsymbol{\beta} = \sum_{j=1}^{p} x_{ij}\beta_j = \eta_i \tag{A.8}$$

where g is a monotone, differentiable function. Also from (3.5)

$$\text{var}(Y_i) = [b''(\theta_i)c'(\theta_i) - c''(\theta_i)b'(\theta_i)]/[b'(\theta_i)]^3 \tag{A.9}$$

The score with respect to parameter β_j is defined as

$$U_j = \frac{\partial l(\boldsymbol{\theta}; \mathbf{y})}{\partial \beta_j} = \sum_{i=1}^{N} \frac{\partial l_i}{\partial \beta_j}$$

where

$$l_i = y_i b(\theta_i) + c(\theta_i) + d(y_i) \tag{A.10}$$

To obtain U_j we use

$$\frac{\partial l_i}{\partial \beta_j} = \frac{\partial l_i}{\partial \theta_i} \frac{\partial \theta_i}{\partial \mu_i} \frac{\partial \mu_i}{\partial \beta_j}$$

By differentiating (A.10) and substituting (A.7)

$$\frac{\partial l_i}{\partial \theta_i} = y_i b'(\theta_i) + c'(\theta_i) = b'(\theta_i)(y_i - \mu_i)$$

By differentiating (A.7) and substituting (A.9)

$$\frac{\partial \mu_i}{\partial \theta_i} = -\frac{c''(\theta_i)}{b'(\theta_i)} + \frac{c'(\theta_i)b''(\theta_i)}{[b'(\theta_i)]^2} = b'(\theta_i)\,\text{var}(Y_i)$$

By differentiating (A.8)

$$\frac{\partial \mu_i}{\partial \beta_j} = \frac{\partial \mu_i}{\partial \eta_i} \frac{\partial \eta_i}{\partial \beta_j} = x_{ij} \frac{\partial \mu_i}{\partial \eta_i}$$

Hence

$$\frac{\partial l_i}{\partial \beta_j} = \frac{\partial l_i}{\partial \theta_i} \frac{\partial \mu_i}{\partial \beta_j} \bigg/ \frac{\partial \mu_i}{\partial \theta_i} = \frac{(y_i - \mu_i)x_{ij}}{\text{var}(Y_i)} \left(\frac{\partial \mu_i}{\partial \eta_i}\right) \tag{A.11}$$

and therefore

$$U_j = \sum_{i=1}^{N} \frac{(y_i - \mu_i)x_{ij}}{\text{var}(Y_i)} \left(\frac{\partial \mu_i}{\partial \eta_i}\right) \tag{A.12}$$

The elements of the information matrix are defined by $\mathcal{J}_{jk} = E(U_j U_k)$. From (A.11), for each Y_i the contribution to $\mathcal{J}_{jk}$ is

$$E\left[\frac{\partial l_i}{\partial \beta_j} \frac{\partial l_i}{\partial \beta_k}\right] = E\left[\frac{(y_i - \mu_i)^2 x_{ij} x_{ik}}{\{\text{var}(Y_i)\}^2} \left(\frac{\partial \mu_i}{\partial \eta_i}\right)^2\right] = \frac{x_{ij} x_{ik}}{\text{var}(Y_i)} \left(\frac{\partial \mu_i}{\partial \eta_i}\right)^2$$

and therefore

$$\mathcal{J}_{jk} = \sum_{i=1}^{N} \frac{x_{ij} x_{ik}}{\text{var}(Y_i)} \left(\frac{\partial \mu_i}{\partial \eta_i}\right)^2 \tag{A.13}$$

Appendix C

Here are several versions of analysis of variance for the two-factor experiment shown in Table 7.5. The responses are

$$\mathbf{y} = [6.8, 6.6, 5.3, 6.1, 7.5, 7.4, 7.2, 6.5, 7.8, 9.1, 8.8, 9.1]^{\mathrm{T}}$$

C.1 CONVENTIONAL PARAMETRIZATIONS WITH SUM-TO-ZERO CONSTRAINTS

C.1.1 Saturated model: $E(Y_{jkl}) = \mu + \alpha_j + \beta_k + (\alpha\beta)_{jk}$

$$\boldsymbol{\beta} = \begin{bmatrix} \mu \\ \alpha_1 \\ \alpha_2 \\ \alpha_3 \\ \beta_1 \\ \beta_2 \\ (\alpha\beta)_{11} \\ (\alpha\beta)_{12} \\ (\alpha\beta)_{21} \\ (\alpha\beta)_{22} \\ (\alpha\beta)_{31} \\ (\alpha\beta)_{32} \end{bmatrix} \qquad \boldsymbol{X} = \begin{bmatrix} 110010100000 \\ 110010100000 \\ 110001010000 \\ 110001010000 \\ 101010001000 \\ 101010001000 \\ 101001000100 \\ 101001000100 \\ 100110000010 \\ 100110000010 \\ 100101000001 \\ 100101000001 \end{bmatrix}$$

$$\boldsymbol{X}^{\mathrm{T}}\mathbf{y} = \begin{bmatrix}]Y_{\cdots} \\ Y_{1\cdot\cdot} \\ Y_{2\cdot\cdot} \\ Y_{3\cdot\cdot} \\ Y_{\cdot 1\cdot} \\ Y_{\cdot 2\cdot} \\ Y_{11\cdot} \\ Y_{12\cdot} \\ Y_{21\cdot} \\ Y_{22\cdot} \\ Y_{31\cdot} \\ Y_{32\cdot} \end{bmatrix} = \begin{bmatrix} 88.2 \\ 24.8 \\ 28.6 \\ 34.8 \\ 45.2 \\ 43.0 \\ 13.4 \\ 11.4 \\ 14.9 \\ 13.7 \\ 16.9 \\ 17.9 \end{bmatrix}$$

The 12×12 design matrix $\boldsymbol{X}$ has six linearly independent rows, so we impose six extra conditions in order to solve the normal equations $\boldsymbol{X}^{\mathrm{T}}\boldsymbol{X}\mathbf{b} = \boldsymbol{X}^{\mathrm{T}}\mathbf{y}$. These conditions are

$$\alpha_1 + \alpha_2 + \alpha_3 = 0 \qquad \beta_1 + \beta_2 = 0$$

$$(\alpha\beta)_{11} + (\alpha\beta)_{12} = 0 \qquad (\alpha\beta)_{21} + (\alpha\beta)_{22} = 0$$

$$(\alpha\beta)_{31} + (\alpha\beta)_{32} = 0 \quad \text{and} \quad (\alpha\beta)_{11} + (\alpha\beta)_{21} + (\alpha\beta)_{31} = 0$$

Hence we obtain

$$\mathbf{b} = [7.35, -1.15, -0.2, 1.35, 0.1833, -0.1833, 0.3167, \\ -0.3167, 0.1167, -0.1167, -0.4333, 0.4333]^{\mathrm{T}}$$

and therefore $\mathbf{b}^{\mathrm{T}}\boldsymbol{X}^{\mathrm{T}}\mathbf{y} = 662.62$.

C.1.2 Additive model: $E(Y_{jkl}) = \mu + \alpha_j + \beta_k$

The design matrix $\boldsymbol{X}$ is obtained by omitting the last six columns from the design matrix given in section C.1.1 and

$$\boldsymbol{\beta} = \begin{bmatrix} \mu \\ \alpha_1 \\ \alpha_2 \\ \alpha_3 \\ \beta_1 \\ \beta_2 \end{bmatrix} \qquad \boldsymbol{X}^{\mathrm{T}}\boldsymbol{X} = \begin{bmatrix} 12 & 4 & 4 & 4 & 6 & 6 \\ 4 & 4 & 0 & 0 & 2 & 2 \\ 4 & 0 & 4 & 0 & 2 & 2 \\ 4 & 0 & 0 & 4 & 2 & 2 \\ 6 & 2 & 2 & 2 & 6 & 0 \\ 6 & 2 & 2 & 2 & 0 & 6 \end{bmatrix}$$

$$\boldsymbol{X}^{\mathrm{T}}\mathbf{y} = \begin{bmatrix} 88.2 \\ 24.8 \\ 28.6 \\ 34.8 \\ 45.2 \\ 43.0 \end{bmatrix}$$

$\boldsymbol{X}^{\mathrm{T}}\boldsymbol{X}$ has four independent rows so we impose the extra conditions $\alpha_1 + \alpha_2 + \alpha_3 = 0$ and $\beta_1 + \beta_2 = 0$ to obtain

$$\mathbf{b} = [7.35, -1.15, -0.2, 1.35, 0.1833, -0.1833]^{\mathrm{T}}$$

and $\mathbf{b}^{\mathrm{T}}\boldsymbol{X}^{\mathrm{T}}\mathbf{y} = 661.4133$.

C.1.3 Model omitting effects of levels of factor B: $E(Y_{jkl}) = \mu + \alpha_j$

The design matrix $\boldsymbol{X}$ is obtained by omitting the last eight columns from the design matrix given in section C.1.1 and

$$\boldsymbol{\beta} = \begin{bmatrix} \mu \\ \alpha_1 \\ \alpha_2 \\ \alpha_3 \end{bmatrix} \quad \boldsymbol{X}^{\mathrm{T}}\boldsymbol{X} = \begin{bmatrix} 12 & 4 & 4 & 4 \\ 4 & 4 & 0 & 0 \\ 4 & 0 & 4 & 0 \\ 4 & 0 & 0 & 4 \end{bmatrix} \quad \boldsymbol{X}^{\mathrm{T}}\mathbf{y} = \begin{bmatrix} 88.2 \\ 24.8 \\ 28.6 \\ 34.8 \end{bmatrix}$$

$\boldsymbol{X}^{\mathrm{T}}\boldsymbol{X}$ has three independent rows so we impose the extra condition $\alpha_1 + \alpha_2 + \alpha_3 = 0$ to obtain

$$\mathbf{b} = [7.35, -1.15, -0.2, 1.35]^{\mathrm{T}} \quad \text{and} \quad \mathbf{b}^{\mathrm{T}}\boldsymbol{X}^{\mathrm{T}}\mathbf{y} = 661.01$$

C.1.4 Model omitting effects of levels of factor A: $E(Y_{jkl}) = \mu + \beta_k$

The design matrix $\boldsymbol{X}$ is given by columns 1, 5 and 6 of the design matrix in section C.1.1 and $\boldsymbol{\beta} = [\mu, \beta_1, \beta_2]^{\mathrm{T}}$, $\boldsymbol{X}^{\mathrm{T}}\boldsymbol{X}$ is a 3×3 matrix with two linearly independent rows so we impose the constraint $\beta_1 + \beta_2 = 0$ to obtain

$$\mathbf{b} = [7.35, 0.1833, -0.1833]^{\mathrm{T}} \quad \text{and} \quad \mathbf{b}^{\mathrm{T}}\boldsymbol{X}^{\mathrm{T}}\mathbf{y} = 648.6733$$

C.1.5 Model with only a mean effect: $E(Y_{jkl}) = \mu$

In this case $\mathbf{b} = [\hat{\mu}] = 7.35$ and $\mathbf{b}^{\mathrm{T}}\boldsymbol{X}^{\mathrm{T}}\mathbf{y} = 648.27$.

C.2 CORNER-POINT PARAMETRIZATIONS

C.2.1 Saturated model: $E(Y_{jkl}) = \mu + \alpha_j + \beta_k + (\alpha\beta)_{jk}$ with

$$\alpha_1 = \beta_1 = (\alpha\beta)_{11} = (\alpha\beta)_{12} = (\alpha\beta)_{21} = (\alpha\beta)_{31} = 0$$

$$\boldsymbol{\beta} = \begin{bmatrix} \mu \\ \alpha_2 \\ \alpha_3 \\ \beta_2 \\ (\alpha\beta)_{22} \\ (\alpha\beta)_{32} \end{bmatrix} \quad \boldsymbol{X} = \begin{bmatrix} 100000 \\ 100000 \\ 100100 \\ 100100 \\ 110000 \\ 110000 \\ 110110 \\ 110110 \\ 101000 \\ 101000 \\ 101101 \\ 101101 \end{bmatrix} \quad \boldsymbol{X}^{\mathrm{T}}\mathbf{y} = \begin{bmatrix} Y_{\ldots} \\ Y_{2..} \\ Y_{3..} \\ Y_{12.} \\ Y_{22.} \\ Y_{32.} \end{bmatrix} = \begin{bmatrix} 88.2 \\ 28.6 \\ 34.8 \\ 43.0 \\ 13.7 \\ 17.9 \end{bmatrix}$$

so

$$X^{\mathrm{T}}X = \begin{bmatrix} 12 & 4 & 4 & 6 & 2 & 2 \\ 4 & 4 & 0 & 2 & 2 & 0 \\ 4 & 0 & 4 & 2 & 0 & 2 \\ 6 & 2 & 2 & 6 & 2 & 2 \\ 2 & 2 & 0 & 2 & 2 & 0 \\ 2 & 0 & 2 & 2 & 0 & 2 \end{bmatrix} \qquad \mathbf{b} = \begin{bmatrix} 6.7 \\ 0.75 \\ 1.75 \\ -1.0 \\ 0.4 \\ 1.5 \end{bmatrix}$$

and $\mathbf{b}^{\mathrm{T}}X^{\mathrm{T}}\mathbf{y} = 662.62$.

C.2.2 Additive model: $E(Y_{jkl}) = \mu + \alpha_j + \beta_k$ with $\alpha_1 = \beta_1 = 0$

The design matrix X is obtained by omitting the last two columns of the design matrix in section C.2.1 and so

$$\boldsymbol{\beta} = \begin{bmatrix} \mu \\ \alpha_2 \\ \alpha_3 \\ \beta_2 \end{bmatrix} \qquad X^{\mathrm{T}}X = \begin{bmatrix} 12 & 4 & 4 & 6 \\ 4 & 4 & 0 & 2 \\ 4 & 0 & 4 & 2 \\ 6 & 2 & 2 & 6 \end{bmatrix} \qquad X^{\mathrm{T}}\mathbf{y} = \begin{bmatrix} 88.2 \\ 28.6 \\ 34.8 \\ 43.0 \end{bmatrix}$$

hence

$$\mathbf{b} = \begin{bmatrix} 6.383 \\ 0.950 \\ 2.500 \\ -0.367 \end{bmatrix}$$

and so $\mathbf{b}^{\mathrm{T}}X^{\mathrm{T}}\mathbf{y} = 661.4133$.

C.2.3 Model omitting effects of levels of factor B: $E(Y_{jkl}) = \mu + \alpha_j$ *with* $\alpha_1 = 0$

The design matrix X is obtained by omitting the last three columns of the design matrix in section C.2.1 and so

$$\boldsymbol{\beta} = \begin{bmatrix} \mu \\ \alpha_2 \\ \alpha_3 \end{bmatrix} \qquad X^{\mathrm{T}}X = \begin{bmatrix} 12 & 4 & 4 \\ 4 & 4 & 0 \\ 4 & 0 & 4 \end{bmatrix} \qquad X^{\mathrm{T}}\mathbf{y} = \begin{bmatrix} 88.2 \\ 28.6 \\ 34.8 \end{bmatrix}$$

hence

$$\mathbf{b} = \begin{bmatrix} 6.20 \\ 0.95 \\ 2.50 \end{bmatrix}$$

and $\mathbf{b}^{\mathrm{T}}X^{\mathrm{T}}\mathbf{y} = 661.01$.

C.2.4 Model omitting effects of levels of factor A: $E(Y_{jkl}) = \mu + \beta_k$ *with* $\beta_1 = 0$

The design matrix X is given by columns 1 and 4 of the design matrix in section C.2.1 and so

$$\boldsymbol{\beta} = \begin{bmatrix} \mu \\ \beta_2 \end{bmatrix} \qquad X^TX = \begin{bmatrix} 12 & 6 \\ 6 & 6 \end{bmatrix} \qquad X^T\mathbf{y} = \begin{bmatrix} 88.2 \\ 43.0 \end{bmatrix}$$

hence

$$\mathbf{b} = \begin{bmatrix} 7.533 \\ -0.367 \end{bmatrix}$$

and $\mathbf{b}^TX^T\mathbf{y} = 648.6733$.

C.2.5 Model with only a mean effect: $E(Y_{jkl}) = \mu$

In this case $\mathbf{b} = [\hat{\mu}] = 7.35$ and $\mathbf{b}^TX^T\mathbf{y} = 648.27$.

C.3 ORTHOGONAL VERSION OBTAINED BY A SPECIAL CHOICE OF DUMMY VARIABLES

C.3.1 Saturated model: $E(Y_{jkl}) = \mu + \alpha_j + \beta_k + (\alpha\beta)_{jk}$ *with*

$$\alpha_1 = \beta_1 = (\alpha\beta)_{11} = (\alpha\beta)_{12} = (\alpha\beta)_{21} = (\alpha\beta)_{31} = 0$$

$$\boldsymbol{\beta} = \begin{bmatrix} \mu \\ \alpha_2 \\ \alpha_3 \\ \beta_2 \\ (\alpha\beta)_{22} \\ (\alpha\beta)_{32} \end{bmatrix} \quad \text{and} \quad X = \begin{bmatrix} 1 & -1 & -1 & -1 & 1 & 1 \\ 1 & -1 & -1 & -1 & 1 & 1 \\ 1 & -1 & -1 & 1 & -1 & -1 \\ 1 & -1 & -1 & 1 & -1 & -1 \\ 1 & 1 & 0 & -1 & -1 & 0 \\ 1 & 1 & 0 & -1 & -1 & 0 \\ 1 & 1 & 0 & 1 & 1 & 0 \\ 1 & 1 & 0 & 1 & 1 & 0 \\ 1 & 0 & 1 & -1 & 0 & -1 \\ 1 & 0 & 1 & -1 & 0 & -1 \\ 1 & 0 & 1 & 1 & 0 & 1 \\ 1 & 0 & 1 & 1 & 0 & 1 \end{bmatrix}$$

where the columns of the design matrix X corresponding to terms $(\alpha\beta)_{jk}$ are the products of columns corresponding to terms α_j and β_k. Thus

$$X^TX = \begin{bmatrix} 12 & 0 & 0 & 0 & 0 & 0 \\ 0 & 8 & 4 & 0 & 0 & 0 \\ 0 & 4 & 8 & 0 & 0 & 0 \\ 0 & 0 & 0 & 12 & 0 & 0 \\ 0 & 0 & 0 & 0 & 8 & 4 \\ 0 & 0 & 0 & 0 & 4 & 8 \end{bmatrix} \qquad X^T\mathbf{y} = \begin{bmatrix} 88.2 \\ 3.8 \\ 10.0 \\ -2.2 \\ 0.8 \\ 3.0 \end{bmatrix}$$

hence

$$\mathbf{b} = \begin{bmatrix} 7.35 \\ -0.2 \\ 1.35 \\ -0.1833 \\ -0.1167 \\ 0.4333 \end{bmatrix}$$

and so $\mathbf{b}^TX^T\mathbf{y} = 662.62$.

C.3.2 Additive model: $E(Y_{jkl}) = \mu + \alpha_k + \beta_k$ *with* $\alpha_l = \beta_l = 0$

The design matrix X is obtained by omitting the last two columns of the design matrix in section C.3.1. By the orthogonality of X, estimates of μ, α_2, α_3 and β_2 are the same as in section C.3.1 and hence $\mathbf{b}^TX^T\mathbf{y} = 661.4133$.

C.3.3 Model omitting effects of levels of factor B: $E(Y_{jkl}) = \mu + \alpha_j$ *with* $\alpha_l = 0$

The design matrix X is obtained by omitting the last three columns of the design matrix in section C.3.1. By the orthogonality of X, estimates of μ, α_2 and α_3 are the same as in section C.3.1 and hence $\mathbf{b}^TX^T\mathbf{y} = 661.01$.

C.3.4 Model omitting effects of levels of factor A: $E(Y_{jkl}) = \mu + \beta_k$ *with* $\beta_l = 0$

As before, the estimates of μ and β_2 are the same as in section C.3.1 and $\mathbf{b}^TX^T\mathbf{y} = 648.6733$.

C.3.5 Model with only a mean effect: $E(Y_{jkl}) = \mu$

As before, $\hat{\mu} = 7.35$ and $\mathbf{b}^TX^T\mathbf{y} = 648.27$.

Appendix D

Here are some log-linear models for three-dimensional contingency tables. This is not a complete list. The models are overparametrized so all the subscripted variables are subject to sum-to-zero or corner-point constraints.

D.1 THREE RESPONSE VARIABLES

The multinomial distribution applies,

$$f(\mathbf{y};\, \boldsymbol{\theta}) = n! \prod_{j=1}^{J} \prod_{k=1}^{K} \prod_{l=1}^{L} \theta_{jkl}^{y_{jkl}} / y_{jkl}!$$

1. The **maximal model** is $E(Y_{jkl}) = n\theta_{jkl}$, i.e.

   $$\eta_{jkl} = \mu + \alpha_j + \beta_k + \gamma_l + (\alpha\beta)_{jk} + (\alpha\gamma)_{jl} + (\beta\gamma)_{kl} + (\alpha\beta\gamma)_{jkl}$$

   which has JKL independent parameters.
2. The **pairwise** or **partial association model** is $E(Y_{jkl}) = n\theta_{jk.}\theta_{j.l}\theta_{.kl}$, i.e.

   $$\eta_{jkl} = \mu + \alpha_j + \beta_k + \gamma_l + (\alpha\beta)_{jk} + (\alpha\gamma)_{jl} + (\beta\gamma)_{kl}$$

   with $JKL - (J-1)(K-1)(L-1)$ independent parameters.
3. The **conditional independence model** in which, at each level of one variable, the other two are independent is, for example,

   $$E(Y_{jkl}) = n\theta_{jk.}\theta_{j.l}\theta_{j..}$$

   i.e.

   $$\eta_{jkl} = \mu + \alpha_j + \beta_k + \gamma_l + (\alpha\beta)_{jk} + (\alpha\gamma)_{jl}$$

   with $J(K + L - 1)$ independent parameters.
4. A model with one variable independent of the other two, for example, $E(Y_{jkl}) = n\theta_{j..}\theta_{.kl}$, i.e.

   $$\eta_{jkl} = \mu + \alpha_j + \beta_k + \gamma_l + (\beta\gamma)_{kl}$$

   with $J + KL - 1$ independent parameters.
5. The **complete independence model** is $E(Y_{jkl}) = n\theta_{j..}\theta_{.k.}\theta_{..l}$, i.e.

$$\eta_{jkl} = \mu + \alpha_j + \beta_k + \gamma_l$$

with $J + K + L - 2$ independent parameters.

6. **Non-comprehensive models** do not involve all variables, for example, $E(Y_{jkl}) = n\theta_{jk.}$, i.e.

$$\eta_{jkl} = \mu + \alpha_j + \beta_k + (\alpha\beta)_{jk}$$

with JK independent parameters.

D.2 TWO RESPONSE VARIABLES AND ONE EXPLANATORY VARIABLE

If the third variable is the fixed explanatory one, the product multinomial distribution is

$$f(\mathbf{y};\boldsymbol{\theta}) = \prod_{l=1}^{L} y_{..l}! \prod_{j=1}^{J}\prod_{k=1}^{K} \theta_{jkl}^{y_{jkl}}/y_{jkl}!$$

and all log-linear models must include the term $\mu + \gamma_l$.

1. The maximal model is $E(Y_{jkl}) = y_{..l}\theta_{jkl}$, i.e.

$$\eta_{jkl} = \mu + \alpha_j + \beta_k + \gamma_l + (\alpha\beta)_{jk} + (\alpha\gamma)_{jl} + (\beta\gamma)_{kl} + (\alpha\beta\gamma)_{jkl}$$

with JKL independent parameters.

2. The model describing independence of the response variables at each level of the explanatory variable is $E(Y_{jkl}) = y_{..l}\theta_{j.l}\theta_{.kl}$, i.e.

$$\eta_{jkl} = \mu + \alpha_j + \beta_k + \gamma_l + (\alpha\gamma)_{jl} + (\beta\gamma)_{kl}$$

with $L(J + K - 1)$ independent parameters.

3. The **homogeneity** model in which the association between the responses is the same at each level of the explanatory variable is $E(Y_{jkl}) = y_{..l}\theta_{jk.}$, i.e.

$$\eta_{jkl} = \mu + \alpha_j + \beta_k + \gamma_l + (\alpha\beta)_{jk}$$

with $JK + L - 1$ independent parameters.

D.3 ONE RESPONSE VARIABLE AND TWO EXPLANATORY VARIABLES

If the first variable is the response the product multinomial distribution is

$$f(\mathbf{y};\boldsymbol{\theta}) = \prod_{k=1}^{K}\prod_{l=1}^{L} y_{.kl}! \prod_{j=1}^{J} \theta_{jkl}^{y_{jkl}}/y_{jkl}!$$

and all log-linear models must include the terms

$$\mu + \beta_k + \gamma_l + (\beta\gamma)_{kl}$$

1. The maximal model is $E(Y_{jkl}) = y_{.kl}\theta_{jkl}$, i.e.

$$\eta_{jkl} = \mu + \alpha_j + \beta_k + \gamma_l + (\alpha\beta)_{jk} + (\alpha\gamma)_{jl} + (\beta\gamma)_{kl} + (\alpha\beta\gamma)_{jkl}$$

   with JKL independent parameters.
2. If the probability distribution is the same for all columns of **each** subtable then $E(Y_{jkl}) = y_{.kl}\theta_{j.l}$, i.e.

$$\eta_{jkl} = \mu + \alpha_j + \beta_k + \gamma_l + (\alpha\gamma)_{jl} + (\beta\gamma)_{kl}$$

   with $L(J + K - L)$ independent parameters.
3. If the probability distribution is the same for all columns of **every** subtable then $E(Y_{jkl}) = y_{.kl}\theta_{j..}$, i.e.

$$\eta_{jkl} = \mu + \alpha_j + \beta_k + \gamma_l + (\beta\gamma)_{kl}$$

   with $KL + J - 1$ independent parameters.

Outline of solutions for selected exercises

CHAPTER 1

1.1

$$\begin{pmatrix} W_1 \\ W_2 \end{pmatrix} \sim N\left[\begin{pmatrix} 5 \\ 2 \end{pmatrix}, \begin{pmatrix} 23 & 2 \\ 2 & 53 \end{pmatrix}\right]$$

1.2 (a) $Y_1^2 \sim \chi_1^2$; (b) $\mathbf{y}^{\mathrm{T}}\mathbf{y} = Y_1^2 + (Y_2 - 3)^2/4 \sim \chi_2^2$; (c) $\mathbf{y}^{\mathrm{T}}\mathbf{V}^{-1}\mathbf{y} = Y_1^2 + Y_2^2/4 \sim \chi^2(2, 9/8)$.

1.3 (a) $\bar{Y} \sim N(\mu, \sigma^2/n)$; (c) $(n-1)S^2/\sigma^2 \sim \chi_{n-1}^2$;
(d) If

$$Z = \frac{\bar{Y} - \mu}{\sigma/\sqrt{n}} \sim N(0, 1) \quad \text{and} \quad U^2 = (n-1)S^2/\sigma^2 \sim \chi_{n-1}^2$$

so that $[S^2/\sigma^2]^{1/2} = [U^2/(n-1)]^{1/2}$, then

$$\frac{\bar{Y} - \mu}{S/\sqrt{n}} = \frac{Z}{[\mathrm{U}^2/(\mathrm{n}-1)]^{1/2}} \sim t_{n-1}$$

CHAPTER 2

2.4 (a) Model 1: $\hat{\mu}_1 = 66.8$, $\hat{\mu}_2 = 65.0$ and $\hat{S}_1 = 1339.6$. Model 0: $\hat{\mu} = 65.9$ and $\hat{S}_0 = 1355.8$. Hence $f = 0.2$ which is not statistically significant so H_0 is not rejected and we cannot conclude that there was a change in weight.
(b) Under H_0, $\mu = 0$ and $\hat{S}_0 = \Sigma\, d_k^2 = 80$. Under $H_1, \hat{\mu} = (1/K)\Sigma\, d_k = \bar{d} = 1.8$ and $\hat{S}_1 = \Sigma\,(d_k - \bar{d})^2 = 47.6$. If the D_k's are independent and all have the same distribution, $N(\mu, \sigma^2)$, then $S_1/\sigma^2 \sim \chi_{k-1}^2$. Also if H_0 is correct $S_0/\sigma^2 \sim \chi_k^2$, or if H_0 is not correct S_0/σ^2 has a non-central χ^2 distribution. Hence $_{1,k-1}$ if H_0 is correct. In this case $(\hat{S}_0 - \hat{S}_1)/[\hat{S}_1/(k-1)] = 6.13$ which is significant when compared with the $F_{1,9}$ distribution, so H_0 is rejected and we conclude that there is a change in weight.
(c) The conclusions are different.
(d) For analysis (a) it is assumed that all the Y_{jk}'s are independent and that $Y_{jk} \sim N(\mu_j, \sigma_y^2)$ for all j and k. For analysis

(b) it is assumed that the D_k's are independent and that $D_k \sim N(\mu_1 - \mu_2, \sigma_d^2)$ for all k. Analysis (b) does not involve assuming that Y_{1k} and Y_{2k} (i.e. 'before' and 'after' weights of the same person) are independent, so it is more appropriate.

2.5

$$\mathbf{y} = \begin{bmatrix} 3.15 \\ 4.85 \\ 6.50 \\ 7.20 \\ 8.25 \\ 13.5 \end{bmatrix} \quad X = \begin{bmatrix} 1 & 1.0 & 1.0^2 \\ 1 & 1.2 & 1.2^2 \\ 1 & 1.4 & 1.4^2 \\ 1 & 1.6 & 1.6^2 \\ 1 & 1.8 & 1.8^2 \\ 1 & 2.0 & 2.0^2 \end{bmatrix}$$

$$\boldsymbol{\beta} = \begin{bmatrix} \beta_0 \\ \beta_1 \\ \beta_2 \end{bmatrix} \quad \mathbf{e} = \begin{bmatrix} e_1 \\ e_2 \\ e_3 \\ e_4 \\ e_5 \\ e_6 \end{bmatrix}$$

2.6

$$\mathbf{y} = \begin{bmatrix} Y_{11} \\ Y_{12} \\ Y_{13} \\ Y_{21} \\ Y_{22} \\ Y_{23} \end{bmatrix} \quad X = \begin{bmatrix} 1 & 1 & 1 & 0 \\ 1 & 1 & 0 & 1 \\ 1 & 1 & -1 & -1 \\ 1 & -1 & 1 & 0 \\ 1 & -1 & 0 & 1 \\ 1 & -1 & -1 & -1 \end{bmatrix}$$

$$\boldsymbol{\beta} = \begin{bmatrix} \mu \\ \alpha_1 \\ \beta_1 \\ \beta_2 \end{bmatrix} \quad \mathbf{e} = \begin{bmatrix} e_{11} \\ e_{12} \\ e_{13} \\ e_{21} \\ e_{22} \\ e_{23} \end{bmatrix}$$

CHAPTER 3

3.1 $a(y) = y$, $b(\theta) = -\theta$, $c(\theta) = \phi \log \theta - \log \Gamma(\phi)$ and $d(y) = (\phi - 1) \log y$. Hence $E(Y) = \phi/\theta$ and $\text{var}(Y) = \phi/\theta^2$.

3.2 (a) $\exp[\log \theta - (\theta + 1) \log y]$;

(b) $\exp[\log \theta - y\theta]$;

(c)

$$\exp\left[\log\binom{y + r - 1}{r - 1} + r \log \theta + y \log (1 - \theta)\right]$$

3.3

$$E(U^2) = E(-U') = \text{var}(U) = \frac{n}{\pi(1 - \pi)}$$

3.5 Omitting the data for $i = 1$ if $\log y_i$ is plotted against $\log i$ the slope of a straight line fitted to the plotted points gives an estimate of θ of approximately 2.

3.6 (a) $f(y_i; \pi_i) = \exp\{y_i[\log \pi_i - \log(1-\pi_i)] + \log(1-\pi_i)\}$;
(e) As the dose, x, increases the probability of death, π, increases from near zero to an asymptotic value of 1.

3.7 Yes; $a(y) = e^{y/\phi}$, $b(\theta) = -e^{-\theta/\phi}$, $c(\theta) = -\log(\phi) - \theta/\phi$, $d(y) = y/\phi$.

3.8 No; although the distribution belongs to the exponential family there is no link function g equal to a linear combination of the β's.

CHAPTER 4

4.1 (a)
$$w_{ii} = [\mu_i]_{\boldsymbol{\beta}=\mathbf{b}} = \exp(b_1 + b_2 x_i)$$
$$z_i = (b_1 + b_2 x_i) + y_i \exp[-(b_1 + b_2 x_i)] - 1$$
(b) and (c) $b_1 = -1.944$, $b_2 = 2.175$.

4.2 $\hat{\beta} = \exp(\bar{y})$.

4.3 Use the exponential distribution and the logarithmic link function (for GLIM: $error g$, $ link *l* $, $scale 1$) to obtain $\hat{\beta}_1 = 8.477$ and $\hat{\beta}_2 = -1.109$. From the plot the model appears to describe the data reasonably well.

4.4 (a)
$$l(\beta; \mathbf{y}) = l(\beta^*; \mathbf{y}) + (\beta - \beta^*)\left[\frac{dl(\beta)}{d\beta}\right]_{\beta=\beta^*} + \frac{1}{2}(\beta - \beta^*)^2\left[\frac{d^2 l(\beta)}{d\beta^2}\right]_{\beta=\beta^*}$$
hence
$$b = b^* - \left[\frac{dl(\beta)}{d\beta}\right]_{\beta=b^*} \Big/ \left[\frac{dl^2(\beta)}{d\beta^2}\right]_{\beta=b^*}$$
(b) $\hat{\boldsymbol{\beta}} = \hat{\boldsymbol{\beta}}^* - \boldsymbol{H}^{-1}\mathbf{U}$.

CHAPTER 5

5.1 (a)
$$\mathscr{I} = \frac{n}{\pi(1-\pi)}$$
(b)
$$\frac{(y - n\pi)^2}{n\pi(1-\pi)}$$

(c) $$2\left[y\log\left(\frac{\hat{\pi}}{\pi}\right) + (n - y)\log\left(\frac{1-\hat{\pi}}{1-\pi}\right)\right] \quad \text{where } \hat{\pi} = \frac{y}{n}$$

(d) $P(\chi_1^2 > 3.84) = 0.05$ can be used for the critical value. (i) Wald/score statistic = 4.44, log-likelihood statistic = 3.07; so the first would suggest rejecting $\pi = 0.1$ and the second would not; (ii) both statistics equal 0 and would not suggest rejecting $\pi = 0.3$; (iii) Wald/score statistic = 1.60, log-likelihood statistic = 1.65; so neither would suggest rejecting $\pi = 0.5$.

5.2 (a)

$$2\left[\sum y_i \log\left(\frac{y_i}{n_i\hat{\pi}}\right) + \sum (n_i - y_i)\log\left(\frac{n_i - y_i}{n_i - n_i\hat{\pi}}\right)\right]$$

where

$$\hat{\pi} = \sum y_i \Big/ \sum n_i$$

(b) $2\Sigma\log(\bar{y}/y_i)$ where $\bar{y} = \Sigma y_i/N$.

5.3 (a) $8.477 \pm 1.96 \times 1.655$;
(b) $\Delta D = 26.282 - 19.457 = 6.826$ which is significant when compared with the distribution χ_1^2 so we may conclude that $\beta_2 \neq 0$, i.e. that high initial white cell count is associated with increased survival time.

5.4 Standardized residuals $(y_{jk} - \hat{\mu}_j)/\hat{\sigma}$ when sorted and plotted against normal scores show no apparent departure from linearity (i.e. the assumption of Normality seems reasonable).

5.5 The residual of the last observation (5, 65) is an obvious outlier.

CHAPTER 6

6.1 (a) For refined sugar $y_1 = 39.6 - 4.91x$, where y_1 is consumption of refined sugar and $x = 1, 2, 3, 4, 5$ or 5.7 for periods 1936–39,. . ., 1976–79 and 1983–86 respectively. Slope = change per 10 years = −4.9064 with standard error of 0.5266, so an approximate 95% confidence interval for the average annual decline in consumption is given by $-4.9064/10 \pm 1.96 \times 0.5266/10$, i.e. (−0.59, −0.39). For sugar consumption in manufactured foods $y_2 = 13.2 + 3.88x$, where y_2 is sugar consumption and x is as above. Slope = 3.8843 with standard error = 0.4395 which gives an approximate 95% confidence interval for the average annual rise in consumption of (0.30, 0.47).
(b) For total sugar $y = 52.8 - 1.02x$. Slope = −1.0221 with standard error = 0.7410 so the data are consistent with the hypothesis that there was no change over time (because $z = -1.0221/0.7410 = -1.38$ and so $p > 0.2$).

6.2 A possible model is $y = 6.63 + 0.361P$ where $y = (\text{yield}/1000)^2$ and P = amount of phosphorus.

6.3

Model	Terms	D	Degrees of freedom	ΔD
6.7	Age + weight + protein	567.66	16	
6.8	Weight + protein	606.02	17	38.36
6.9	Age + protein	833.57	17	
6.10	Protein	858.65	18	25.08

Using models (6.7) and (6.8)

$$f = \frac{38.36}{1} \bigg/ \frac{567.66}{16} = 1.08$$

Using models (6.9) and (6.10)

$$f = \frac{25.08}{1} \bigg/ \frac{833.57}{17} = 0.51$$

In this case neither comparison provides evidence against the null hypothesis that response is unrelated to age. More importantly, however, this example shows that analyses to examine the effect of any variable on the response depend on which other explanatory variables are included in the model (unless the variables are orthogonal).

6.4 (c)

Model	D	Degrees of freedom
Age + bmi	26.571	27
Age	31.636	28

To test the effect of body mass index (bmi), after adjustment for age, use

$$f = \frac{31.636 - 26.571}{28 - 27} \bigg/ \frac{26.571}{27} = 5.147$$

which is significant compared with the $F_{1,27}$ distribution. So these data suggest that cholesterol level is positively associated with body mass index.

CHAPTER 7

7.1 (a)

Source of variation	Degrees of freedom	Sum of squares	Mean square	f	p
Mean	1	350.919			
Between groups	2	7.808	3.904	11.65	<0.001
Residual	28	9.383	0.335		
Total	31	368.110			

Compared with the $F_{2,28}$ distribution the value of $f = 11.65$ is very significant so we conclude the group means are not all equal. Further analyses are needed to find which means differ.

(b) (−0.098, 1.114) indicating that the means for the obese groups do not differ significantly.

(c) the residuals show some tendency to increase with increasing plasma phosphate levels.

7.2

Source of variation	Degrees of freedom	Sum of squares	Mean square	f	p
Mean	1	51122.50			
Between workers	3	54.62	18.21	14.45	<0.001
Between days	1	6.08	6.08	4.83	<0.05
Interaction	3	2.96	0.99	0.79	
Residual	32	40.20	1.26		
Total	40	51226.36			

There are significant differences between workers and between days but no evidence of interaction effects.

7.3

Model	Deviance	Degrees of freedom
$\mu + \alpha_j + \beta_k + (\alpha\beta)_{jk}$	5.00	4
$\mu + \alpha_j + \beta_k$	6.07	6
$\mu + \alpha_j$	8.75	7
$\mu + \beta_k$	24.33	8
μ	26.00	9

(a)
$$f = \frac{6.07 - 5}{2} \bigg/ \frac{5}{4} = 0.43$$

so there is no evidence of interaction;
(b) (i) $\Delta D = 18.26$; (ii) $\Delta D = 17.25$. The data are unbalanced so the model effects are not orthogonal.

7.4

Model	Deviance	Degrees of freedom
$\mu_j + \alpha_j x$	9.63	15
$\mu_j + \alpha x$	10.30	17
$\mu + \alpha x$	27.23	19
μ_j	26.86	18
μ	63.81	20

(a)
$$f = \frac{63.81 - 26.86}{2} \bigg/ \frac{26.86}{18} = 12.38$$

which indicates that the treatment effects are significantly different, if the initial aptitude is ignored;
(b)
$$f = \frac{10.30 - 9.63}{2} \bigg/ \frac{9.63}{15} = 0.52$$

so there is no evidence that initial aptitude has different effects for different treatment groups.

CHAPTER 8

8.1 A good model is $\text{logit}\,\pi = -3.458 + 0.0066\exp(\text{dose})$, where π is the probability that a cancer death is due to leukaemia and dose = 1, 2, . . ., 6 for radiation dose = 0, 1–9, . . ., 200+ (better models can be obtained by defining the dose as the lower limit of each dose interval, 0, 1, 10 . . .).

8.2 (a) $\phi = \exp(\beta_1 - \beta_2) = 1$ if and only if $\beta_1 = \beta_2$;
(b) $\phi_j = \exp[(\alpha_1 - \alpha_2) + x_j(\beta_1 - \beta_2)]$ is constant if $\beta_1 = \beta_2$.

8.3 Proportions of schoolboys planning to attend college increased with IQ and SES. The following model describes the data well: $\text{logit}\,\pi_{jk} = \mu + \alpha_j + \beta_k$, where α_j and β_k denote parameters for SES and IQ respectively, ignoring the ordering among the categories. For corner-point parametrization with $\alpha_1 = 0$ and $\beta_1 = 0$ the maximum likelihood estimates for α_j and β_k are shown below (with standard errors in brackets).

Parameter	Estimate (s.e.)
Base (SES:L, IQ:L)	−2.93 (0.12)
SES: LM	0.60 (0.11)
UM	1.16 (0.11)
H	2.13 (0.11)
IQ: LM	0.82 (0.11)
UM	1.62 (0.11)
H	2.39 (0.11)

8.4 (c) $a = b = \frac{1}{2}$, $\mathrm{var}[\psi(t)] = (1 - \pi)/n\pi$.

8.5

	Logistic model	Linear model
a_1	0.877	0.724
a_2	1.284	0.825
b	−0.155	−0.040
D	2.619	2.472

The fitted values are very similar for both models.

CHAPTER 9

9.1 $X^2 = 17.648$ and $D = 18.643$, both with 2 degrees of freedom, so the significance level is less than 0.001 and we conclude that responses differ for the placebo and vaccine groups. The table of standardized residuals is

	Response		
	Small	Medium	Large
Placebo	2.21	−1.50	−1.15
Vaccine	−2.30	1.57	1.20

This shows that the vaccine produces higher levels of immune response.

9.2 (a) Satisfaction was highest in the tower blocks and lowest in houses; contact was highest in houses and lowest in tower blocks; satisfaction was higher when contact was higher and, in particular, satisfaction was highest in tower blocks with high contact.
(b) There was strong evidence of association between satisfaction and type of housing ($D = 34.53$, degrees of freedom $= 2$, $p < 0.001$), between contact and type of housing ($D = 39.06$, degrees of freedom $= 2$, $p < 0.001$) and, once these two effects are included in the model, there is also evidence of association between satisfaction and contact ($D = 8.87$, degrees of freedom $= 2$, $p < 0.005$).

9.4 (c) Take aspirin use as the response Z with the binomial distribution $b(n, \pi)$, where n is the total number of users and non-users. The probability π is modelled by logit (π) with site as a main effect, corresponding to (9.12), or site and case/control status as main effects, corresponding to (9.13). The resulting log-likelihood ratio values are the same as for the log-linear models.

References

Aitkin, M., Anderson, D., Francis, B. and Hinde, J. (1989) *Statistical Modelling in GLIM*, Clarendon Press, Oxford.

Andersen, E. B. (1980) *Discrete Statistical Models with Social Science Applications*, North-Holland, Amsterdam.

Barndorff-Nielsen, O. (1978) *Information and Exponential Families in Statistical Theory*, Wiley, New York.

Belsley, D. A., Kuh, E. and Welsch, R. E. (1980) *Regression Diagnostics: Identifying Influential Data and Sources of Collinearity*, Wiley, New York.

Berkson, J. (1953) A statistically precise and relatively simple method of estimating the bio-assay with quantal response, based on the logistic function. *J. Amer. Statist. Assoc.*, **48**, 565–99.

Birch, M. W. (1963) Maximum likelihood in three-way contingency tables. *J. R. Statist. Soc. B*, **25**, 220–33.

Bishop, Y. M. M., Fienberg, S. E. and Holland, P. W. (1975) *Discrete Multivariate Analysis: Theory and Practice*, MIT Press, Cambridge, Mass.

Bliss, C. I. (1935) The calculation of the dosage-mortality curve. *Annals of Applied Biology*, **22**, 134–67.

Chambers, J. M. (1973) Fitting non-linear models: numerical techniques. *Biometrika*, **60**, 1–13.

Charnes, A., Frome, E. L. and Yu, P. L. (1976) The equivalence of generalized least squares and maximum likelihood estimates in the exponential family. *J. Amer. Statist. Assoc.*, **71**, 169–71.

Cook, R. D. and Weisberg, S. (1982) *Residuals and Influence in Regression*, Chapman and Hall, London.

Cox, D. R. and Snell E. J. (1989) *Analysis of Binary Data*, 2nd edn, Chapman and Hall, London.

Cox, D. R. and Hinkley, D. V. (1974) *Theoretical Statistics*, Chapman and Hall, London.

Cox, D. R. and Snell, E. J. (1968) A general definition of residuals. *J. R. Statist. Soc. B*, **30**, 248–75.

Cox, D. R. and Snell, E. J. (1981) *Applied Statistics: Principles and Examples*, Chapman and Hall, London.

Cressie, N. and Read, T. R. C. (1989) Pearson's χ^2 and the loglikelihood ratio statistic G^2: a comparative review. *Inter. Statistical Rev.*, **57**, 19–43.

Draper, N. R. and Smith, H. (1981) *Applied Regression Analysis*, 2nd edn, Wiley, New York.

Duggan, J. M., Dobson, A. J., Johnson, H. and Fahey, P. P. (1986) Peptic ulcer and non-steroidal anti-inflammatory agents. *Gut*, **27**, 929–33.

Everitt, B. S. (1977) *The Analysis of Contingency Tables*, Chapman and Hall, London.

Fahrmeir, L. and Kaufman, H. (1985) Consistency and asymptotic normality of the maximum likelihood estimator in generalized linear models. *Annals of*

Statistics, **13**, 342–68.
Fienberg, S. E. (1980) *The Analysis of Cross-Classified Categorical Data*, 2nd edn, MIT Press, Cambridge, Mass.
Finney, D. J. (1973) *Statistical Method in Biological Assay*, 2nd edn, Hafner, New York.
Fox, D. R. (1986) MINITAB as a teaching aid for generalized linear models. In *Pacific Statistical Congress* (eds I. S. Francis *et al.*) Elsevier, Amsterdam, pp. 317–20.
Freeman, D. H., Jr (1987) *Applied Categorical Data Analysis*, Marcel Dekker, New York.
Grizzle J. E., Starmer, C. F. and Koch, G. G. (1969) Analysis of categorical data by linear models. *Biometrics*, **25**, 489–504.
Healy, M. J. R. (1988) *GLIM: An Introduction*, Clarendon Press, Oxford.
Hocking, R. R. (1985) *The Analysis of Linear Models*, Brooks/Cole, Monterey.
Holliday, R. (1960) Plant population and crop yield. *Field Crop Abstracts*, **13**, 159–67, 247–54.
Jones, R. H. (1987) Serial correlation in unbalanced mixed models. *Bull. International Statistical Institute*, **52**, book 4, 105–22.
Kleinbaum, D. G. Kupper, L. L. and Muller, K. E. (1988) *Applied Regression Analysis and other Multivariable Methods*, 2nd edn, P.W.S.-Kent, Boston, Mass.
McCullagh, P. (1980) Regression models for ordinal data. *J. R. Statist. Soc. B*, **42**, 109–42.
McCullagh, P. and Nelder, J. A. (1986) *Generalized Linear Models*, 2nd edn, Chapman and Hall, London.
McKinlay, S. M. (1978) The effect of nonzero second-order interaction on combined estimators of the odds ratio. *Biometrika*, **65**, 191–202.
Madsen, M. (1976) Statistical analysis of multiple contingency tables. Two examples. *Scand. J. Statist.*, **3**, 97–106.
NAG (Numerical Algorithms Group) (1985) *The GLIM System Release 3.77 Manual* (ed. C. D. Payne), NAG, Oxford.
Nelder, J. A. (1974) Log linear models for contingency tables: a generalization of classical least squares. *Appl. Statist.*, **23**, 323–9.
Nelder, J. A. and Wedderburn, R. W. M. (1972) Generalised linear models. *J. R. Statist. Soc. A*, **135**, 370–84.
Otake, M. (1979) *Comparison of Time Risks Based on a Multinomial Logistic Response Model in Longitudinal Studies*, Technical Report No. 5, RERF, Hiroshima, Japan.
Pierce, D. A. and Schafer, D. W. (1986) Residuals in generalized linear models. *J. Amer. Statist. Assoc.*, **81**, 977–86.
Pregibon, D. (1981) Logistic regression diagnostics. *Annals of Statist.*, **9**, 705–24.
Ratkowsky, D. A. (1983) *Nonlinear Regression Modeling*, Marcel Dekker, New York.
Ratkowsky, D. A. and Dolby, G. R. (1975) Taylor series linearization and scoring for parameters in nonlinear regression. *Appl. Statist.*, **24**, 109–11.
Roberts, G., Martyn, A. L., Dobson, A. J. and McCarthy, W. H. (1981) Tumour thickness and histological type in malignant melanoma in New South Wales, Australia, 1970–76. *Pathology*, **13**, 763–70.
Ryan, B. F., Joiner, B. L. and Ryan, T. A., Jr (1985) *MINITAB Handbook*, 2nd edn, Duxbury, Boston.

Sangwan-Norrell, B. S. (1977) Androgenic stimulating factors in the anther and isolated pollen grain culture of *Datura innoxia* Mill. *J. Expt. Botany,* **28**, 843–52.

Sewell, W. H. and Shah, V. P. (1968) Social class, parental encouragement and educational aspirations. *Amer. J. Sociol.*, **73**, 559–72.

Sinclair, D. F. and Probert, M. E. (1986) A fertilizer response model for a mixed pasture system. In *Pacific Statistical Congress* (eds I. S. Francis *et al.*), Elsevier, Amsterdam, pp. 470–74.

Walter, S. D., Feinstein, A. R. and Wells, C. K. (1987) Coding ordinal independent variables in multiple regression analyses. *Am. J. Epidemiol.*, **125**, 319–23.

Whittaker, J. and Aitkin, M. (1978) A flexible strategy for fitting complex log-linear models. *Biometrics,* **34**, 487–95.

Whyte, B. M., Gold, J., Dobson, A. J. and Cooper, D. A. (1987) Epidemiology of acquired immunodeficiency syndrome in Australia. *Med. J. Aust.*, **146**, 65–9.

Winer, B. J. (1971) *Statistical Principles in Experimental Design,* 2nd edn, McGraw-Hill, New York.

Wood, C. L. (1978) Comparison of linear trends in binomial proportions. *Biometrics,* **34**, 496–504.

Index